T0387934

Electromagnetism and the Metonymic Imagination

Published in collaboration with the Society for Literature, Science, and the Arts, AnthropoScene presents books that examine relationships and points of intersection among the natural, biological, and applied sciences and the literary, visual, and performing arts. Books in the series promote new kinds of cross-disciplinary thinking arising from the idea that humans are changing the planet and its environments in radical and irreversible ways.

Electromagnetism and the Metonymic Imagination

Kieran M. Murphy

The Pennsylvania State
University Press
University Park,
Pennsylvania

Library of Congress Cataloging-in-Publication Data

Names: Murphy, Kieran M. (Kieran Marcellin), 1977– author.
Title: Electromagnetism and the metonymic imagination / Kieran M. Murphy.
Other titles: AnthropoScene.
Description: University Park, Pennsylvania : The Pennsylvania State University Press, [2020] | Series: AnthropoScene: the SLSA book series | Includes bibliographical references and index.
Summary: "Illustrates how the discovery of electromagnetism in 1820 not only led to technological inventions, such as the dynamo and the telegraph, but also legitimized modes of reasoning that manifested a sharper ability to perceive how metonymic relations could reveal the order of things"—Provided by publisher.
Identifiers: LCCN 2019058663 | ISBN 9780271086057 (hardcover)
Subjects: LCSH: Electromagnetism in literature. | Poe, Edgar Allan, 1809–1849—Criticism and interpretation. | Balzac, Honoré de, 1799–1850—Criticism and interpretation. | Villiers de L'Isle-Adam, Auguste, comte de, 1838–1889—Criticism and interpretation. | Metonyms. | Literature and science.
Classification: LCC PN56.E49 M87 2020 | DDC 809/.9336—dc23
LC record available at https://lccn.loc.gov/2019058663

Printed in the United States of America
Published by The Pennsylvania State University Press,
University Park, PA 16802-1003

The Pennsylvania State University Press is a member of the Association of University Presses.

It is the policy of The Pennsylvania State University Press to use acid-free paper. Publications on uncoated stock satisfy the minimum requirements of American National Standard for Information Sciences—Permanence of Paper for Printed Library Material, ANZI Z39.48-1992.

And who shall calculate the immense influence upon social life—upon arts—upon commerce—upon literature—which will be the immediate result of the great principles of electro-magnetics?

Edgar Allan Poe, “The Man That Was Used Up” (1839)

Contents

Acknowledgments

Extraordinary scholars helped bring this book to completion. I would like to express my gratitude to Sydney Lévy for his insight and friendship. Heartfelt thanks to Laurence Rickels, Catherine Nesci, Eric Prieto, Chris Braider, Warren Motte, Paul Youngquist, Elise Arnould-Bloomfield, Paul Harris, Ed Dimendberg, and Linda Dalrymple Henderson for their suggestions and encouragement over the years.

Many thanks to editor Kendra Boileau and her team and to series editors Lucinda Cole and Bob Markley for seeing this project through. The book has greatly benefited from their feedback and attentive care as well as from the critical engagement of their anonymous readers.

I am grateful to the journals *SubStance* and *Épistémocritique* for publishing the first articles where I began to sharpen the ideas contained in this book. I further developed them thanks to two volumes edited by Aura Heydenreich and Klaus Mecke (De Gruyter) and by Victoire Feuillebois and Émilie Pezard (Minard), respectively.

I received generous support from various institutions at the University of Colorado-Boulder, where I was awarded a faculty fellowship from the Center for Humanities and the Arts and received funding for archival research from the Center for the Study of Origins and the Graduate Committee on the Arts and Humanities.

Support also came in indirect yet important ways. I would like to thank especially Lynn A. Higgins, J. Kathleen Wine, Laure Marcellesi, Håkan P. Tell, Steve Kurtz, Michael Minelli, David L. Feinberg, and on a more personal note, Nelly, Geneviève, Brendan, Catherine, Garry, Jason, Niall, Candy, John, and Andrew.

Finally, I would like to dedicate this book to my first readers: Annie, Garret, and Aimee.

Introduction

When we invoke "the electric age," we should understand this expression as actually referring to the electro*magnetic* age. By omitting "magnetic," we lose track of the discoveries that transformed daily life and the global economy. In 1820 Hans Christian Oersted noticed that a current-carrying wire deflected a nearby compass needle (fig. 1). The detection of a mysterious link between electricity and magnetism took the European scientific community by surprise. Most believed that these natural forces were unrelated. Oersted's discovery showed otherwise and led to the creation of a new field of research, electromagnetism, prompted by the need to study the two forces in tandem.

Electromagnetism quickly became the subject of an intense investigation that culminated in 1831, when Michael Faraday uncovered another striking effect known as "induction." He showed that he could generate an electric current in a conductor by simply moving a magnet near it. Electromagnetic induction made possible several world-changing technologies, such as the telegraph and the dynamo, and the latter was responsible for the advent of mass electrification during the second half of the nineteenth century.

The study of electromagnetism presented scientists with a new set of challenges that forced them not only to revise their conception of reality but also to think differently about how to investigate it. To illustrate the profound epistemological shift instigated by electromagnetism during the nineteenth century, Paul Valéry devised a thought experiment inspired by the phenomenon of induction at work in the dynamo. If we had access to the place where the great pre–nineteenth-century minds reside (Valéry jokingly calls this place "hell"), and were to give a dynamo to Archimedes, Galileo, Descartes, or Newton, they would not know what to do with it. They would spin the movable part, take the device apart, measure all its pieces, and never have a chance to penetrate its secret. The dynamo would confound them because, for Valéry, these luminaries could only think of "mechanical transformations."[1] Such thinking could

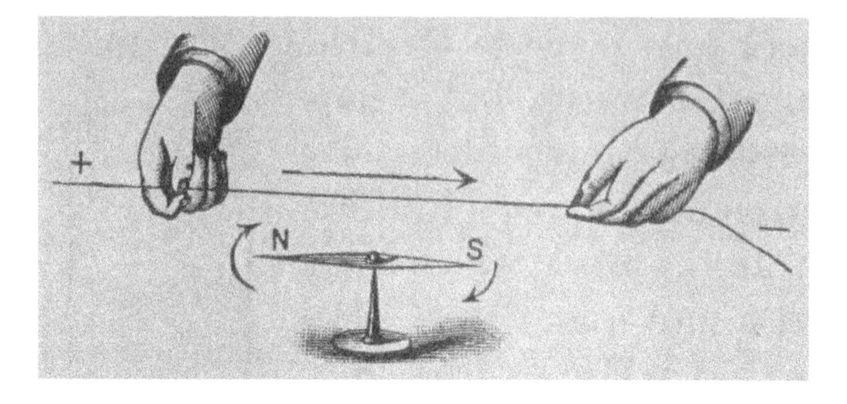

Fig. 1 Oersted's groundbreaking experiment, which revealed that an electric current deflects a nearby magnetic needle. Henry S. Carhart and Horatio N. Chute, *Practical Physics* (Boston: Allyn and Bacon, 1920), 366, figure 381. Library of Congress, Washington, DC.

grasp machines such as the classical clock, lever, or balance but could not have understood the dynamo.[2]

Electromagnetic induction manifests a transformation of energy that eludes mechanical interpretation. The great achievements accumulated by Newtonian physics since the eighteenth century legitimized this mechanical worldview associated with industrialization and the flowering of bourgeois society during the nineteenth century. However, a significant component of the world continued to elude the Newtonian conception of nature. The application of Newtonian physics to electromagnetism yielded only limited results. The classical laws of electromagnetism, what we now call "Maxwell's equations," took form in 1861 thanks to the elaboration of a new physical concept that originated in the patterns formed by iron filings around a magnet.

Faraday called these patterns "lines of force" and thought that they represented electric and magnetic forces better than the Newtonian law of universal gravitation, which states that gravitational attraction occurs between separate objects without delay or apparent mediation. Faraday thought that the instantaneous force at work in Newton's atomist universe seemed too magical to be true and saw in the curved lines of force extending beyond the magnet a more accurate way to represent a force acting in space. Whereas space is absolute and plays no part in gravitational attraction, lines of force show that a magnet reconfigures the surrounding space to move iron filings or to attract other magnets.

The concept of a magnetic field implies that objects are not really separate. They participate in the same continuum characterized by a malleable and active space that contains their potential for action. Now known as "field theory," this new representation of the transmission of energy broke away from Newtonian atomism and prompted James Clerk Maxwell to develop a mathematical formalism compatible with a continuous physical reality. Besides the laws of electromagnetism, Faraday's lines of force helped Maxwell formulate the electromagnetic wave theory from which he predicted the existence of the radio wave and identified light as an electromagnetic phenomenon. In 1905 Einstein published his special theory of relativity, which builds on field theory and an asymmetry in Maxwell's interpretation of electromagnetic induction to show that time and space are indeed malleable. Einstein's special theory of relativity completed Faraday's critique of Newtonian physics and ended a theoretical system that had dominated the intellectual climate of the previous one hundred and fifty years.

In the dynamo thought experiment, Valéry contrasts mechanical with electromagnetic transformations to highlight a different kind of analytical thinking closely connected to this profound reconceptualization of nature. What made this alternative thinking so radically different was that it took root in electromagnetic phenomena. In the 1820s, alongside the mechanical thinking that characterized classical luminaries, a different way to explore and order our surroundings began to emerge that broadly could be called "electromagnetic thinking."

This book examines some of the key elements in electromagnetic thinking that helped make the secret of the dynamo intelligible. These elements are inseparable from the apparatuses—what I call transformational motors and Romantic machines—that helped materialize and legitimize them. I therefore organize this study around three types of apparatuses powered by electromagnetism: chains, the lab experiment Faraday used to unveil the phenomenon of induction, and automata. My aim is not to provide an exhaustive historical account of these objects. Through reading strategies drawn from literary studies and the field of science and literature, I concentrate instead on how they bolstered the emergence of electromagnetic thinking.

The legacy of electromagnetism in the history of science and technology has recently been the subject of important revisions and clarifications by, among others, David Gooding, Ryan D. Tweney, Christine Blondel, Friedrich Steinle, Kenneth Caneva, and Françoise Balibar. Yet beyond physics treatises and laboratories, lesser-known intellectual legacies also deserve attention. Electromagnetism supported ideas and practices based on an interpretation of reality more organic and interconnected than previous worldviews. Like many scientific theories, it soon captured the attention and imagination of humanists and social reformers. I argue that literature became a site of textual experimentation that engaged with early interpretations of electromagnetism and prompted significant changes in the understanding of language, social relations, and polarities such as subject and object, mind and nature, conscious and nonconscious, and life and death. I will show how these changes occurred as evidenced by analogies inspired by the newly found link connecting electricity and magnetism, and I will demonstrate how the images these analogies produced helped authors explore and redefine social, aesthetic, epistemological, and metaphysical domains. This overlooked dialogue between science and literature provides a new perspective on critical debates that shaped the nineteenth century.

In recent years electricity has been the subject of growing interest in literary studies.[3] Sam Halliday has followed the emergence of a kind of "electrical thinking" in works by Nathaniel Hawthorne, Herman Melville, Mark Twain, and Henry James, where electrical communication technologies such as the telegraph and the telephone provided key models to reimagine not only interpersonal connections but the nature of thinking itself. Paul Gilmore has examined the impact of electricity on modern definitions of aesthetics and has shown how, from British Romantics to the American Renaissance and from Edmund Burke to Frederick Douglass, it rearticulated and complicated the relationship between the aesthetic and political spheres. Jennifer Lieberman has demonstrated how Mark Twain, Charlotte Perkins Gilman, Jack London, Ralph Ellison, and Lewis Mumford relied on electrical technologies to undermine dominant modes of thinking associated with American industrialization and individualism and to promote alternative visions of social interconnectedness. Many contemporary discussions of networks and systems can be traced back to scientific and literary discourses addressing the nature of electricity.

My study contributes to this scholarship by shifting the focus from electricity to electromagnetism.[4] Authors have traditionally relied on either electricity *or* magnetism to convey the nature of human interactions such as power relations and romantic attraction. After the discovery of electromagnetism, these interactions could also be described through the link between electricity *and* magnetism. I will show that these remarkable electromagnetic analogies appeared as early as 1833 and examine how they impacted the works of three main authors. I trace their emergence in the writings of Honoré de Balzac and

Edgar Allan Poe and examine their legacies in Villiers de l'Isle-Adam. I employ a comparative approach to recognize patterns and structures that extend beyond individual, national, and genre specificities. Comparing examples across time helps clarify the origin and function of electromagnetic analogies and their profound cultural and epistemological impacts.

These analogies shed light not only on the history of literature but also on scientific thinking. One of the central aims of the humanistic field of science and literature, according to Devin Griffiths, "is to explain the role of imaginative language in science and to explore the impact of literary form on scientific practice."[5] Hans Christian Oersted's 1820 detection of a relation between the electric current and magnetism resulted in great part from a research program that took its cue from the "polarity" and "unity" of natural forces, heavily contested notions disseminated by Romantic literature and *Naturphilosophie*.[6] By unifying the domains of electricity and magnetism, his discovery provided empirical evidence for these notions. It also contributed, I argue, to the rehabilitation of the discourse of analogy.

Griffiths characterizes the nineteenth century as "the age of analogy." His study of the interactions between science and literature in the works of Erasmus and Charles Darwin shows that writers and scientists increasingly relied on various types of analogies involving comparisons, tropes, and "correspondences" to challenge boundaries and drive conceptual innovation. Such analogical approaches had fallen into disrepute due to their association with pre-Enlightenment theological, philosophical, and alchemical methods. Yet they continued to exist as alternative modes of analytical thinking in the margins of the intellectual establishment. I contend that the unexpected identification of a link between electricity and magnetism provided crucial empirical grounds to relegitimate and redefine the discourse of analogy. I examine the appearance of electromagnetic tropes in literature and show how they anticipated similar analogies in scientific and philosophical discourses.

My analysis demonstrates that electromagnetic analogies provided a more complex model of reality that called into question prevalent views concerning how things relate through space and time. Whereas electric imagery tended to emphasize metaphorical relations founded on resemblance, electromagnetic imagery underscored metonymic relations based on contiguity. In Poe's little-known tale "The Spectacles," which I discuss in chapter 1, the narrator describes how love "at first sight" invisibly links bodies across space not just through its similarity to an electric connection but also through a relation of contiguity between electricity and magnetism. Oersted's and Faraday's experiments with electromagnetic interaction proved that the two forces are contiguous phenomena that share an intimate link despite their spatial separation and different physical properties. In "The Spectacles," the representation of falling in love does not simply depend on an electric "fluid" *or* magnetic fluid anymore; it is an interaction between the two. The reality portrayed in such electromagnetic imagery manifests a shift from metaphoric to metonymic relations where spatial and temporal separation are not simply subsumed by the undifferentiated continuity of a single "fluid."

The metonymic shift that I trace in the complex models unveiled by nineteenth-century electromagnetic analogies prefigures the explosion of metonymic reasoning that marked the beginning of the twentieth century, which Ronald Schleifer has identified in influential notions such as Walter Benjamin's "constellation," Mikhail Bakhtin's "interfacings of 'borders,'" Bertrand Russell's "arrangement of order," Albert Einstein's "operational definition," and Werner Heisenberg's "alternation."[7] This earlier metonymic shift is therefore significant because it contributes to our understanding of not only the history of analogical thinking but also some of the less familiar origins of modern metonymic reasoning. Electromagnetic thinking also sheds light on a critical turning point that contributed to rediscovering modes of thought based on relations of contiguity.

Metonymic Contiguity and the Imagination

Building upon the ideas of Roman Jakobson, George Lakoff, and Mark Johnson, cognitive linguists have recently obtained greater empirical evidence that metonymy is not merely a figure of speech but a fundamental cognitive process that plays a central role in the way we produce and order meaning, interpretation, and knowledge.[8] Electromagnetism has greatly contributed to the revalorization of metonymic reasoning, often working as a model for its conceptualization. For instance, when Jakobson formulated his theory on the bipolar structure of language, he relied on a model informed by field theory.[9] According to Jakobson, metaphor and metonymy are the two fundamental "poles" that structure all linguistic systems. Despite their marked difference, they are inseparable. Linguistic formations always depend on a combination of the two that emphasizes one or the other.

Sebastian Matzner recently reevaluated Jakobson's theory of the bipolar structure of language from a tropological perspective and produced more support for the claim that metaphor and metonymy are the two fundamental tropes from which all the others can be derived.[10] He also pointed out that definitions of metonymy as the trope of "association," "proximity," "propinquity," or, in the general modern usage, "contiguity" have been particularly vague concerning the actual relation they supposedly describe. According to philosophy and psychology, contiguity implies "a relation based solely on frequently experienced togetherness, without the necessary involvement of *any* logical principal as such." Yet this more contingent aspect of contiguity has been downplayed by rhetoricians who, since antiquity, have attempted to define metonymy through ever-expanding categories of relations that have usually foregrounded the substitutive logic of synecdoche: "place and inhabitant, individual and group, producer and product, container and contained, cause and effect, and so forth."[11] This emphasis has caused problems of classification, often leading to theories considering synecdoche as distinct from metonymy.[12] The lack of clarity concerning the exact nature of its operation also translated into a neglect of metonymy, which, compared with metaphor, has received less critical attention.

Matzner contends that metonymy, as the trope of contiguity, should represent logical and nonlogical relations. Its general principle should account for synecdoche and more contingent links based on "frequently experienced togetherness" such as "heart" and "courage." Like Jakobson, he turns to a type of field theory—the linguistic concept of "semantic fields"—to reformulate this principle on more all-encompassing grounds.

Before field theory, earlier scientific interpretations of (electro-)magnetic phenomena had already made important contributions to the understanding of the elusive concept of metonymic contiguity. Jakobson's use of metaphor and metonymy is particularly indebted to associationist theories,[13] which elevated resemblance and contiguity to central operations of the mind and which have been linked to electricity and magnetism since David Hume. The philosopher claims that his greatest contribution to the "science of man" was to recast its foundation in the operations of "the imagination." According to Hume, the function of the imagination is to order the mass of impressions and ideas crowding the mind through association. This ability to make connections is the primary engine of the imagination, which produces "chains of thought" that can lead to both "reverie" and scientific insight.[14]

For Hume, the principles of association come down to three basic types of connections that occur through resemblance, contiguity in time and space, and cause and effect. He describes this fundamental ability to make associations as "a kind of ATTRACTION, which in the mental world will be found to have as extraordinary effects as in the natural, and to show itself in as many and as various forms."[15] Hume compares his scientific approach to that of Newton by transposing the latter's use of the term *attraction* to the domain of the mind. Although Newton is famous for his work on gravity, his use of the term refers to all natural forces, including electricity and magnetism.[16] Hume's influential elevation of the imagination and its associative processes based on resemblance and contiguity (or what Jakobson and cognitive linguists call metaphor and metonymy) already depended on an analogy informed by electric and magnetic attractions to materialize how the imagination—and more broadly, thought—works.

As the understanding of these two forces changed during the first half of the nineteenth century, so did thought. As electromagnetic phenomena contributed to the rehabilitation of the discourse of analogy, cognitive association based on contiguity received fresh empirical backing and new conceptual options for representing its mode of operation. In the same movement, these new options redefined the nature of relation and difference, providing critical tools to rethink the interconnection of things in more metonymic terms.

The way this redefinition shaped nineteenth-century literature and science has been obscured due to the rise to prominence of Einstein's theory of relativity and its reconceptualization of electromagnetism. Throughout her works, Linda Dalrymple Henderson has reminded us that, up until 1919, cultural productions were not impacted by Einstein's physics and were instead informed by previous interpretations of electromagnetism. This book traces the legacies of these lesser-known models of electromagnetism dating back to the 1830s.

Imaging (Electro-)Magnetic Contiguity

To clarify how electromagnetic phenomena have expressed and shaped relations of contiguity, I will turn to their visual representation as a kind of chain in Faraday's and Maxwell's diagrams. These images participate in an older and influential tradition that, since Plato's *Ion*, relied on the magnetic chain to visualize metonymic as well as metaphoric relations. A brief discussion of this tradition will reveal an overlooked yet crucial cultural context informing these scientific diagrams and their significance for the concept of contiguity.

In *Ion*, Socrates relies on an analogy inspired by a magnetic chain to convince the rhapsodist Ion that his famed poetic declamation depends on an art devoid of technical or rational skill. The strong influence his performance has on spectators derives instead from its participation in a larger series of connections emanating from a divine source:

> The gift which you possess of speaking excellently about Homer is not a technique, but, as I was just saying, an inspiration; there is a divinity moving you, like that contained in the stone which Euripides calls a magnet, but which is commonly known as the stone of Heraclea. This stone not only attracts iron rings, but also imparts to them a similar power of attracting other rings; and sometimes you may see a number of pieces of iron and rings suspended from one another so as to form quite a long chain: and all of them derive their power of suspension from the original stone. In like manner the Muse first of all inspires men herself; and from these inspired persons a chain of other persons is suspended, who take the enthusiasm.[17]

This passage anticipates Plato's more pronounced opposition between the ideal forms and their lowly copies in the *Republic*. Although rhapsodic interpretation constitutes an embodiment of the divine, it is only the result of a chain of duplications. Such interpretation happens in the enthusiastic state provoked by divine inspiration or possession (*éntheon*). The poet is out of his mind, and as if in a Bacchic trance, he becomes the medium of the Muse who invested him.

The Platonic opposition between divine ideal forms and their mundane copies has been an ongoing influence on the critical discourse on art. It often privileges philosophy and science at the expense of art, since mimesis can only aspire to resemble a model that will always remain out of reach or more truthful. Yet as Jean-Luc Nancy's deconstructive reading of *Ion* has demonstrated, the magnetic chain analogy also manifests a metonymic relation that undermines this emphasis on the metaphorical relation linking art to its model and, consequently, the hierarchical separation of scientific and artistic disciplines.[18] When the poet is in a state of "enthusiasm," gods do the talking. Nancy translates this transmission of the divine voice (*theîa moîra*) as "*le partage divin*" (divine sharing) and "*le partage des voix*" (sharing voices). In French, the word *partage* means both participating, as in sharing, and dividing, as in cutting. This double logic expresses something paradoxical about Ion's art that magnetism also renders apparent.

The divine *partage* is like magnetization. The iron rings participate in the overflowing power of the lodestone while remaining separate from the source of their attraction. Correspondingly, the chain of "inspired persons" comes together not because they share a resemblance with the divine but due to an elusive connection made apparent by their proximity. In Plato, divine magnetization depends on a link marked by difference that relates seemingly unrelated domains through metonymic contiguity. This concatenation occurs through a process of *partage*, which involves a combination of metonymic as well as metaphoric relations and which does not necessarily degrade art to mere imitation. The initial model of the Platonic chain in *Ion* depends on an analogy with magnetic power that is instrumental in making its complex vision of the interconnection of things cling together.

Ion marks the beginning of a long and influential tradition that has represented and explored the inner workings of the cosmos through notions of magnetism. As Koen Vermeir demonstrates, analogies with the magnetic chain and, more broadly, the lodestone's ability to attract and magnetize iron are found in early philosophical and theological texts to support various conceptions of the order and continuity of things. This magnetic imagery appears in works as wide-ranging as the *Corpus Hermeticum* (second century C.E. and later) and the writings of the church fathers. Augustine directly refers to the magnetic chain as an object of fascination: "Who would not be amazed at this virtue of the stone, subsisting as it does not only in itself, but transmitted through so many suspended rings, and binding them together by invisible links?"[19] Through the Middle Ages and the modern era, magnetism remained a popular model to render manifest divine attraction and the "invisible links" connecting God and "His" creation.

Such magnetic "invisible links" inspired a striking iconography that proliferated in the seventeenth century. In 1600 William Gilbert's discovery of geomagnetism provided a scientific foundation for visions of the cosmos as inherently magnetic. Supporters of this "magnetic philosophy" included luminaries such as Johannes Kepler and Robert Boyle. Magnetic philosophy was an important contributor to the scientific, theological, and artistic productions of this era. Its most famous proponent was the Jesuit polymath Athanasius Kircher.

He published several books on magnetic philosophy, two of which contained remarkable frontispieces depicting magnetic chains. They show the hand of God holding the end of magnetic chains that descend from the heavens to Earth and that, along the way, connect with various objects and bigger rings with images inside of them. In Kircher's *Magnes sive de arte magnetica opus tripartitum* (1641), the chains pass through symbols of the Holy Roman Empire, highlighting the political ramifications of Kircher's magnetic vision of the cosmos (fig. 2). The rings also frame images representing various disciplines, which are part of a greater interacting whole connecting the mundane and divine worlds and which range from theology, philosophy, poesis, rhetoric, and music to cosmography, mechanics, astronomy, arithmetic, natural magic, and medicine.[20] In *Magneticum Naturae Regnum* (1667), the images depict various symbols of "sympathies" associated with magnetism at the time (fig. 3). Both frontispieces have banners proclaiming that everything is linked by "arcane knots" (*arcanis nodis*).

Contiguous rings bonded together by magnetic force give form to these "arcane knots."[21] Evoking the description in *Ion*, the magnetic chain offers a remarkable empirical model to make sense of difference and relation within a vision of the cosmos where everything is not only metaphorically but also metonymically interconnected. God is a divine lodestone from whom a transformative power emanates. Just as the lodestone turns iron into a magnet, this power transforms everything it passes through. The magnetic chain brings support to a conception of God as absolutely different, yet everything can partake in "His" power. The contiguity of the rings conveys this "invisible link" or "arcane knot" between the divine and the mundane by representing a relation marked by difference. God's concatenation stays whole thanks to a metonymic power associated with magnetism.

The Platonic tradition of the magnetic chain helped pave the way for the metonymic shift that occurred in the wake of electromagnetism during the nineteenth century. To support this claim, I will turn to Faraday's and Maxwell's use of diagrams. These diagrams were visual cognitive tools that, along with experiments and mathematical reasoning, played a crucial role in their elaboration of the classic laws of electromagnetism. Faraday and Maxwell explored electromagnetic phenomena through images that allowed them to visualize and synthesize particular aspects of the puzzling interaction of electricity and magnetism. David Gooding has shown how Faraday's diagrams built on each other to generate new insights and guide his line of inquiry as he

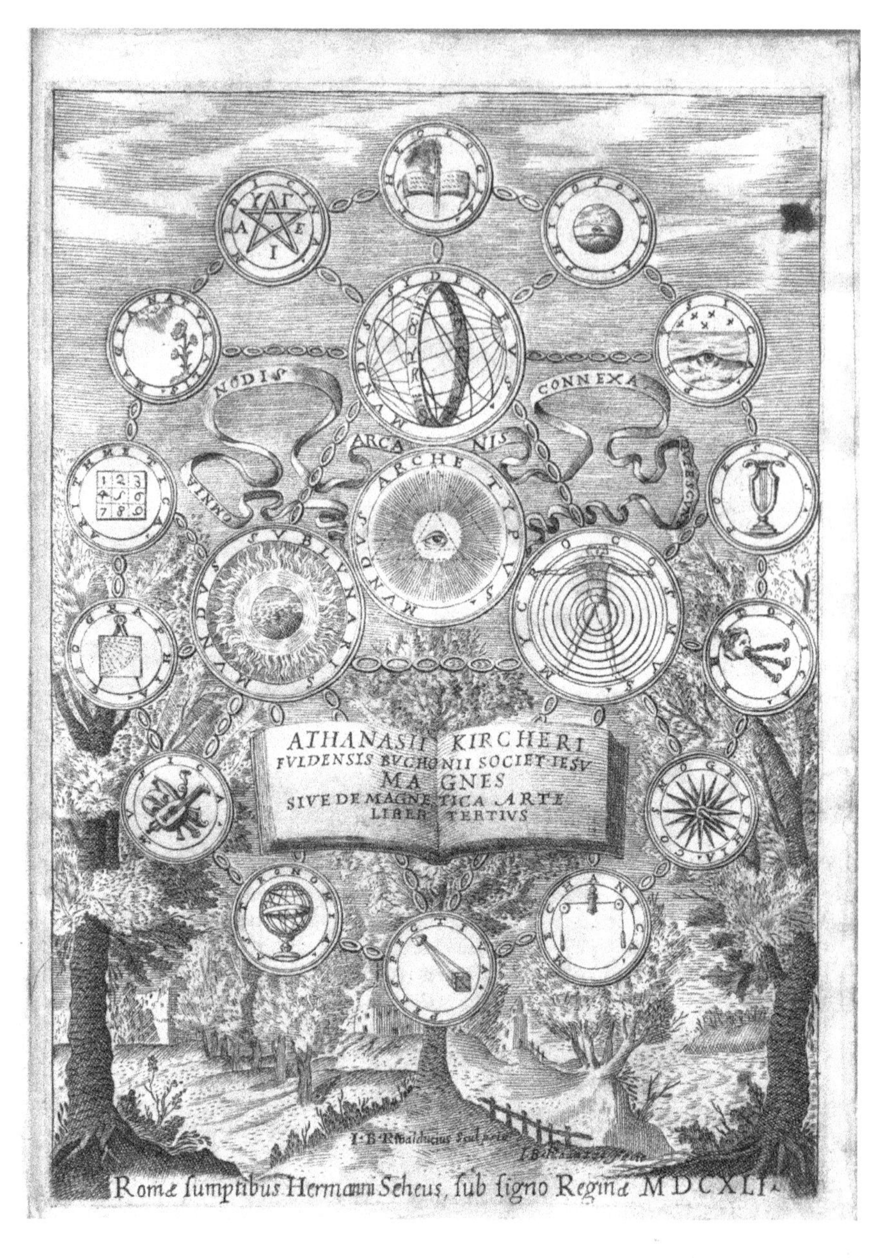

Fig. 2 Frontispiece from Athanasius Kircher's *Magnes sive de arte magnetica opus tripartitum* (1641). Courtesy of the Department of Special Collections, Stanford University Libraries, RBC QC751 K58 1641.

Fig. 3 Frontispiece from Athanasius Kircher's *Magneticum Naturae Regnum* (1667). Courtesy of the Department of Special Collections, Stanford University Libraries, RBC Q155 K58 1667 t.

developed the theory of magnetic lines of force and established the foundation of field theory.[22] According to Thomas K. Simpson, the diagrams accompanying Maxwell's theoretical work on electromagnetism were "figures of thought," which provided effective mathematical and physical analogies for the elusive nature of the field.[23]

Gooding and Simpson have examined these diagrams as they appear within the confines of Faraday's and Maxwell's works. Their close readings do not attend to their relation to the wider visual and cultural context that informed them. Beyond its caption, the meaning of a diagram depends on older and more familiar iconography that shapes the context of its perception, providing a recognizable background on which its author can highlight the contribution he is making. The visual trope of the magnetic chain offers such familiar background and helps clarify the conceptual innovations expressed by some of Faraday's and Maxwell's diagrams. Before turning to these visual representations of electromagnetism and metonymic contiguity, I will describe in more detail the groundbreaking line of thought that they helped formalize.

In the 1850s, following years of meticulous experimentation, Faraday became confident that "magnetic curves," or what he renamed "lines of force," provided an effective physical model to measure and conceptualize the nature of electromagnetic phenomena. He observed that magnetic force exerts its power along curved lines of force that could be rendered visible through the patterns produced by iron filings near a magnet (fig. 4).

The lines of force clearly extend beyond the actual magnet, providing a way to visualize the medium responsible for the propagation of action through space. Faraday thought them mathematically useful because the trajectory they followed between magnetic poles represented the direction of the force at any given point in space. They also provide an effective way to calculate the magnitude of the force. As the patterns of iron filings show, lines of force are closer together near a magnetic pole, where the force is stronger, than further away from it. The different concentration of lines of force allows the measurement of the force's magnitude anywhere in the magnetic sphere of action.[24]

Faraday thought that his idea of lines of force could provide an alternative to Newtonian action at a distance. Although Newton himself did not believe that gravity could operate in a vacuum, his law of gravitational attraction ($F = Gm1m2/r^2$) implied that its action took place between two masses (m1 and m2) instantaneously through space as if without mediation.[25] Space remains unaltered or absolute in Newtonian action at a distance. Faraday believed lines of force around magnetic bodies modified the configuration of space around them and that such modifications are responsible for attraction, thereby laying the foundation of field theory.

The shift from action at a distance to magnetic lines of force provided a radically new theoretical framework that played a crucial role in uncovering the general principles of puzzling phenomena such as the induction of an electrical current by a magnet in motion. In 1852 Faraday synthesized most of his findings concerning such electromagnetic interaction in a diagram that became one of the starting points of Maxwell's mathematical theory of electromagnetism (fig. 5).[26] This image represents the interconnection of electric and magnetic action. Faraday schematizes this quantitative relation in terms

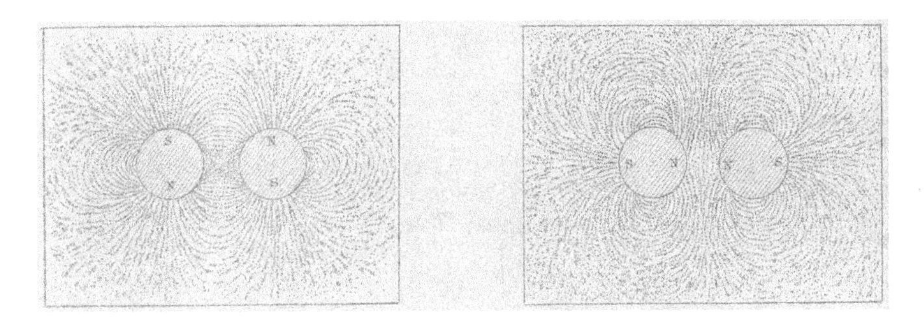

Fig. 4 Lines of magnetic force diagrams from Faraday's *Experimental Researches in Electricity, Volume* 3 (London: R. and J. E. Taylor, 1855), plate IV (detail). Library of Congress, Washington, DC.

of two rings representing electricity (E) and magnetism (M), respectively. The invisible link responsible for this equivalence takes the form of a traditional chain of two interlocking rings. Yet the rings are not phenomenologically the same thing, and their interaction does not derive from a simple mechanical exertion of one ring against the other. It derives from a relation of contiguity. Like the Platonic magnetic chain, their invisible link is represented by a gap, which recalls Kircher's contiguous rings but in this case is much wider. This larger degree of separation visually accentuates the metonymic nature of electromagnetic interaction.

In the 1860s, Faraday's idea of lines of force led Maxwell to synthesize and cast into a set of related equations everything known at the time about electromagnetism.[27] Today the laws of electromagnetism are referred to as "Maxwell's equations," a slightly modified version of his original work.[28] Beyond grand synthesis, Maxwell also predicted from his equations the existence of radio waves. He showed that covarying and mutually inducing electric and magnetic fields exhibited over time a wavelike pattern that could self-propagate in free space. In 1888 Heinrich Hertz announced that he had detected the existence of radio waves, thereby validating Maxwell's self-propagating electromagnetic wave theory. The discovery of self-propagating electromagnetic waves ushered in the age of wireless telecommunication and the revolutionary impact that radio, television, and cell phone would have on the twentieth century.

Maxwell's electromagnetic laws also led to the unification of optics and electromagnetism. He calculated the speed of his theoretical self-propagating electromagnetic wave and found that it was nearly the same as the speed of light. At the time, physicists envisioned light as a wavelike disturbance or vibration

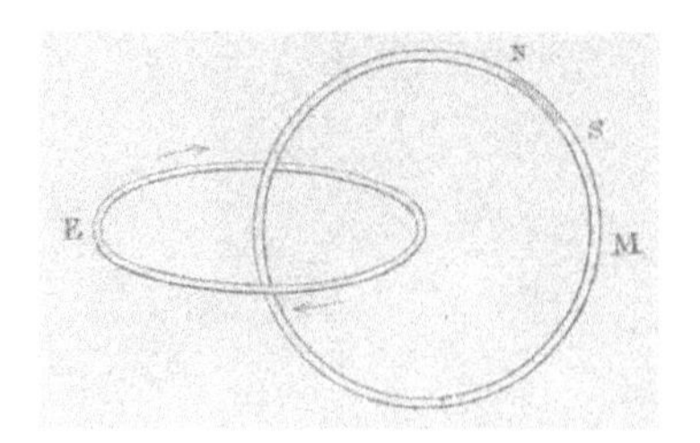

Fig. 5 Diagram from Faraday's *Experimental Researches in Electricity, Volume 3* (London: R. and J. E. Taylor, 1855), plate IV (detail). Library of Congress, Washington, DC.

transmitted through its own ubiquitous medium called the "luminiferous ether." From these striking similarities, and a known link between light and magnetism uncovered by Faraday ("magneto-optic rotation," or the "Faraday effect"), Maxwell inferred that the luminiferous ether was in fact the electromagnetic medium and that light was a type of electromagnetic wave.[29] Phenomena associated with light such as infrared (heat) and ultraviolet radiation also turned out to be electromagnetic waves. Furthermore, detection of radio waves by Hertz foreshadowed the discovery of other types of electromagnetic radiation such as X-rays in 1895 and uranium's radioactivity in 1896. By the end of the nineteenth century, Maxwell's electromagnetic theory of light had become widely accepted, and scientists agreed that wide-ranging phenomena such as radio waves, heat, visible light, ultraviolet, X-rays, and radioactivity shared the same electromagnetic fabric. Their distinct behavior derives from their different wavelengths.

In one of the most groundbreaking and influential diagrams of nineteenth-century theoretical physics (fig. 6), Maxwell conveys the wavelike pattern of mutually inducing electric and magnetic fields and its similarity to light.[30] He depicts this pattern through related electric and magnetic fluctuations occurring on separate perpendicular planes. The planes are contiguous, coming into contact along an axis that marks their separation as well as connection. The vertical orientation of the diagram and the ringlike shapes of electric and magnetic parabolas recall Kircher's magnetic chains descending from the heavens. Electromagnetism is now a self-propagating chain, where the metonymic power holding it together does not come from a divine source anymore. It derives from Maxwell's sophisticated equations and the way they establish links between the electric and magnetic domains through quantitative relation.

In Faraday's and Maxwell's diagrams the Platonic model of the magnetic chain has become an electromagnetic motor. Instead of a divine lodestone, concatenation and metonymic connection result from a physical interaction between electricity and magnetism. This interaction hinges on a newly found

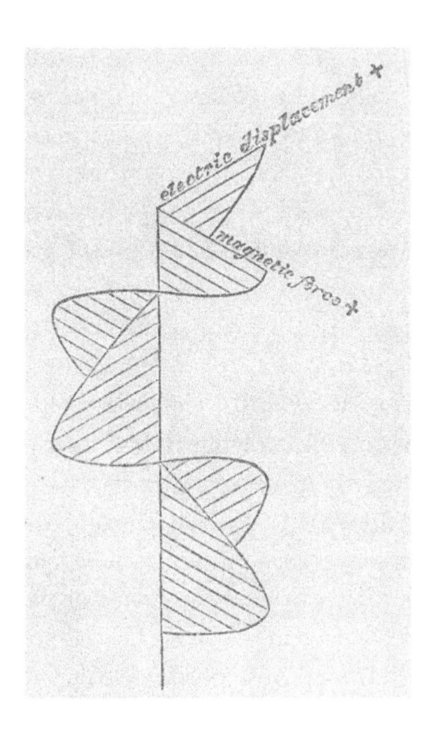

Fig. 6 Diagram from Maxwell's *A Treatise on Electricity and Magnetism*, vol. 2 (Oxford: Clarendon Press, 1873), 390. Library of Congress, Washington, DC.

electromagnetic difference and relation that helped nineteenth-century scientific thinking move beyond divine and Newtonian frameworks.

(Electro-)Magnetism

The depiction of electromagnetism as a chain shows not only the emergence of a new way to express metonymic contiguity but also its close association with previous magnetic models and their visual representation. Yet the legacy of magnetism in the rise of electromagnetic thinking has been overshadowed by many recent critical studies of electricity. Throughout this book, I shift the focus from electricity to electromagnetism by paying closer attention to the less familiar pole of the "electric age"—magnetism.

Graeme Gooday situates the turning point in the omission of magnetism at the influential 1881 International Exposition of Electricity in Paris. Publicists adopted the more convenient but less precise term "electrical" to describe the electromagnetic technologies that were transforming the world.[31] Compared to magnetism, electricity was a relative newcomer to science. It became prominent during the eighteenth century due to Benjamin Franklin's lightning experiment and the inventions of the Leyden jar and voltaic pile.[32] Magnetic properties have been well known since antiquity. Furthermore, the arrival of the magnetic compass in Europe during the Middle Ages had a profound impact on history.[33] By 1620, Francis Bacon famously listed it along with the printing press and gunpowder as the three main discoveries responsible for the making of the modern world.[34]

Magnets have been popular models for exploring and representing various types of motors or unmoved movers of divine or human origin. Before Plato described poetic "enthusiasm" and mimesis as magnetic, the pre-Socratic philosopher Thales of Miletus had already relied on the lodestone to describe the nature of the soul. Another fundamental property of magnetism known since the thirteenth century, bipolarity, provided the impetus to make sense of puzzling polarities such as attraction and repulsion, love and hate, and mind and matter. Faraday saw in magnetic "lines of force" a radical new model to rethink the nature of gravitational as well as electrical interactions, paving the way for Maxwell's and Einstein's elaboration of field theory. As a tool for conceptual exploration and innovation, magnetic analogies illuminate the cultural and epistemological origins of the electromagnetic age.

Consequently, I attend closely to the transition from magnetic to electromagnetic models. Early electromagnetic analogies often represented various metonymic relations traditionally described through categories of magnetization and magnetic bipolarity. For many writers and philosophers, magnetic phenomena helped explore other transformative powers and explained how opposites could attract or be intimately related. Examining the transition from magnetic to electromagnetic concepts and tropes provides an effective way to contextualize what was at stake when Oersted's and Faraday's discoveries began to reorder the world in the lab and beyond.

Transformational Motors and Romantic Machines

Faraday's and Maxwell's diagrams represent a metonymic power generating motion that derives not from a divine source but from an electromagnetic interaction. Through their mutual induction, electricity and magnetism move things. In a dynamo, the transformation of motion into an electric current is reversible. This current can activate magnetic objects, such as the engine of a streetcar. This new way to produce motion participates in the rise of what Michel Serres calls "transformational motors" and associates with the invention of the steam engine. This new type of machine changed not only the world but also how we understand it.[35]

Along with transformational motors came a redefinition of the origin of motion that played a fundamental role in the epistemological shift characterizing nineteenth-century cultural and scientific production. From Aristotle's "unmoved mover" to the neoclassic period, the ultimate cause of all motion in the universe remained purely metaphysical. Ancient motors such as a spring or a water mill relayed the motive force provided by human, animal, or natural actions, which themselves worked as relays of the primordial motor. The steam engine did not simply transport and transmit movement; it appeared to generate its own motive force by transforming heat into mechanical work. This remarkable motor turned age-old metaphysical inquiries concerning the origin of movement into a physical problem. In 1824 the founder of thermodynamics, Sadi Carnot, began to provide a scientific explanation to this problem when he demonstrated that the motive force of the steam engine depended on a temperature difference between hot and cold sources—namely, between the furnace and the condenser.[36] Carnot claimed a temperature difference displaces the metaphysical motor as the source of movement.

Beyond mines, factories, and locomotives, the steam engine embodied a shift from the metaphysical to the secular production of motion transpiring concurrently in the sciences, arts, and humanities. Serres has traced how influential figures such as Hegel, Turner, Darwin, Marx, Zola, Nietzsche, and Freud attempt to seize the means of production of their respective subject matters by displacing metaphysical intervention with the generative power of difference. Their wide-ranging works not only rely on analogies inspired by the steam engine; they themselves function as transformational motors.

Literary scholars such as Bruce Clarke, Barri Gold, and Sydney Lévy have also relied on the steam engine and thermodynamics as their main guide to examine the relation between artistic and scientific changes that occurred during the early period of industrialization.[37] Yet the attention commanded by the steam engine has overshadowed the importance and impact of other transformational motors, such as electromagnetism. This book argues that the electromagnetic motor powered the next and insufficiently understood cycle of the industrial revolution and much of its significant cultural production.[38]

It unveiled a new kind of difference—rendered manifest by the metonymic relation between electricity and magnetism—that became, like the temperature difference in the steam engine, another key source of engineered movement. By the mid-nineteenth century, electromagnetic difference was transforming the world through its implementation in the telegraph and, by the end of the century, through dynamos and power plants that brought forth the first wave of mass electrification.

One of the main challenges in better understanding the transformational power of electromagnetic difference consists in recovering the points of intersection among wide-ranging disciplines and practices where its influence initially spread and thrived. Situated at the interstices of literature and science, these points of convergence generate a fresh perspective on how electromagnetic difference became one of the main transformational motors that changed the world as it empowered analogical methods based on relations of contiguity.

Electromagnetism can be understood as one of the transformational motors that John Tresch has recently called "Romantic machines."[39] The steam engine and the electromagnetic motor are Romantic machines because they could symbolize Romantic as well as mechanical interpretations of the world. Although the two perspectives have traditionally been depicted as incompatible, they shared essential characteristics. As motors of industrialization, they advanced the mechanization of society and the power of clockwork rationalism. As volatile and seemingly self-propelled machines based on striking conversion processes, they also embodied prominent ideas of Romanticism: organicism, metamorphosis, and the unity of natural forces, reason and imagination, and mind and nature.

Romantic machines provided a common ground for mechanical and Romantic aspirations where new conceptions of knowledge production that Tresch broadly labels "mechanical Romanticism" flourished. The hybrid epistemologies of mechanical Romanticism contrasted with the enlightenment ideal of detached objectivity by emphasizing the interdependence of the perceiving subject and the perceived object in the establishment of facts and truths. Influential mechanical Romantics such as Saint-Simon, François Arago, and Auguste Comte produced theories of knowledge that unified scientific, artistic, and spiritual domains in concerted efforts to achieve social and political transformations.

Tresch traces the rise and fall of mechanical Romanticism during the first half of the nineteenth century in Paris. Unlike the exponential specialization and fragmentation of knowledge production that followed in its wake, mechanical Romanticism thrived on interactions among artists, scientists, philosophers, and social and political reformers. The development of realist and fantastic literature, thermodynamics, electromagnetism, socialism, and ecology are some of the wide-ranging yet related outcomes that resulted from such interactions.

Whereas the steam engine has all but disappeared from contemporary life, electromagnetic machines remain ubiquitous and will continue to shape our behavior and the environment for a long time to come. Tresch concentrates on only one case study involving electromagnetism, the work and life of André-Marie Ampère, the other great pioneer of electromagnetic science. Ampère built electromagnetic machines that, soon after Oersted's discovery, helped mathematically prove the equivalence of the electric current and magnetism and establish the principles of electrodynamics. Ampère's achievements were an important influence on Faraday and Maxwell, and the latter famously characterized his French predecessor as the "Newton of electricity."[40] As Tresch shows, Ampère was not simply a mechanical thinker. His breakthrough investigation of electromagnetism was integral to his larger Romantic interests, which included open-ended scientific methods, how the mind interacts with the material world, the unity of natural forces and knowledge, and "animal magnetism."

The present book is entirely dedicated to electromagnetism. It focuses on how its discovery contributed to the development of Romantic machines and helped materialize and legitimize metonymic reasoning. It is structured around three types of apparatuses powered by electromagnetism: chains, the lab experiment Faraday used to unveil the phenomenon of induction, and automata. In the first part, I trace the emergence of a metonymic shift in early nineteenth-century conceptions of interconnection through the apparition of electromagnetic chains in Poe's oeuvre. From his mesmeric to detective tales, Poe relies on the relation of contiguity these chains rendered manifest to undermine traditional interpretations of the great chain of being and its ordering of things based on metaphoric gradation.

These electromagnetic chains provide empirical support for his nonlinear and ironic vision of physical and metaphysical interconnections by drawing extensively on popular theories concerning animal magnetism. These theories attempted to make sense of "magnetic" somnambulists and their strange states of dissociation. Their split-selves unveiled relations of contiguity within the mind that, many believed, were linked to death, haunting, and mourning. I show that Poe accentuates the contiguous nature of these phenomena by not simply relying on the older mesmeric idea of a magnetic fluid that connects bodies through space and time. Instead, he drew inspiration from newer views that had incorporated some of the metonymic implications of Oersted, Ampère, and Faraday's discoveries and that I call "animal electromagnetism."

In the second part, I contend that Balzac pioneered the literary exploration of animal electromagnetism and its metonymic thinking. Anticipating Valéry's dynamo thought experiment, he found in Faraday's induction apparatus a new model to represent relations of contiguity, which he initially invoked to cast inductive reasoning under a new light. His reinterpretation of the method most closely associated with scientific thinking sharply contrasted with the myth of the detached scientist. His rapprochement of electromagnetic and scientific inductions conveys a process of discovery where metonymic reasoning and the body play integral roles.

I also show how his references to electromagnetism inform the realist framework of his novel and enable him to conceive an alternative understanding of space and time intimately related to Einstein's later redefinition of the spatiotemporal fabric of the universe. Although these two luminaries produced completely different works, their respective theories were the products of a similar engagement with the relation of contiguity unveiled by Faraday's induction experiment. Faraday's new, transformational motor rendered manifest a puzzling relation and difference between electricity and magnetism that both Balzac and Einstein mobilized as an engine of conceptual exploration and innovation that, I argue, has been subsequently overshadowed by the rise to prominence of field theory.

The groundbreaking model of electromagnetic contiguity that emerges in Balzac's and Poe's writings is less known because other analogies inspired by the telegraph, telephone, and dynamo quickly became more popular during the time separating their works from Einstein's. Yet during the second half of the nineteenth century, this model continued to be influential through hybrid analogies where electromagnetic interaction and the new technologies it made possible appeared together. In chapter 3, I examine some of the legacies of these hybrid analogies through the remarkable account of an electromagnetic automaton in Villiers de L'Isle-Adam's *L'Ève future* (*Tomorrow's Eve*, 1878–86). This complex novel has been considered by critics and historians to be an important contribution to the philosophical tradition of conceptualizing the nature of life and cognition in terms of self-moving machines. To reach a better grasp of how Villiers's automaton signals epistemological changes concerning these perennial notions, I argue that it needs to be resituated within its (electro-) magnetic context.

The way this automaton derives its power from electromagnetic induction instead of a spring or steam engine is significant. It participates in an older and influential line of thought that culminated in *Naturphilosophie* and that had relied on self-moving machines propelled by magnetic bipolarity (such as the compass) to identify what made the universe move. For Schelling and Goethe, "Nature" was a mysterious, Romantic machine where conflicting elements such as life and death or mind and matter shared a relation of contiguity, which became intelligible through analogy with the opposite yet related poles of a magnet. In Villiers's automaton, the poles now establishing this relation of contiguity were electricity and magnetism. Through their interaction, they provided the conceptual motor that allowed the novel to explore the metonymic nature of life and cognition and that anticipated the invention of the unconscious.

I conclude with an extended discussion where I connect Balzac's and Poe's trailblazing use of electromagnetic contiguity with the broader explosion of metonymic reasoning that marked the first half of the twentieth century. I focus on the theoretical writings of philosopher Gaston Bachelard and writer Julien Gracq. Bachelard knew about the instrumental role Faraday's induction experiment played in Einstein's discovery, and he employed its metonymic dimension to elaborate his seminal idea of "epistemological break" and its nonlinear conception of the history of science. Although his philosophical and literary works have often been criticized for their lack of correspondence, the role of electromagnetic induction in Bachelard's oeuvre shows that it was a unifying concept that helped him represent the rise of a "new spirit" in literature as well as science.

Building upon Bachelard's literary theory, Julien Gracq wrote an important yet understudied essay on André Breton where he inaugurated the critical exploration of the *electromagnetic* imagination. For Gracq, electromagnetic induction lent its coherence to the avant-garde aesthetics of surrealism because it sprung from practices like automatic writing, where chance encounter through association based on contiguity played a defining role. I provide translations and close readings of key passages in the essay to show how, much like Poe and Balzac before him, Gracq mobilized electromagnetic induction to conceive of a mode of metonymic communication that, unlike the mere duplication of experience found in mimetic poetry, could overcome the constraints of linguistic mediation.

1.

(Electro-)Magnetic Chains

Poe's "The Spectacles," or The Insight of Short Sight

Through ironic detachment, Poe's little-known tale "The Spectacles" (1844) brings into focus the mid-nineteenth-century fascination with everything electromagnetic.[1] As it humorously exploits popular interest in Morse's electromagnetic telegraph and Mesmer's "animal magnetism," it brings to the fore the metonymic power of love at first sight. It does so in the tale's opening lines, when Poe's narrator gives an unprecedented electromagnetic account of the nature of love:

> Many years ago, it was the fashion to ridicule the idea of "love at first sight;" but those who think, not less than those who feel deeply, have always advocated its existence. Modern discoveries, indeed, in what may be termed ethical magnetism or magnetœsthetics, render it probable that the most natural, and, consequently, the truest and most intense of the human affections are those which arise in the heart as if by electric sympathy—in a word, that the brightest and most enduring of the psychal fetters are those which are riveted by a glance. The confession I am about to make will add another to the already almost innumerable instances of the truth of the position.[2]

Poe's narrator is young, romantic, and shortsighted. Too vain to wear eyeglasses, he unknowingly fell in love at first sight with his French great-great-grandmother and is reminiscing about the embarrassing events that led to their mock marriage. He recalls that he fell in love with her at the opera after perceiving her "most exquisite figure" in a far-off private box. He chased after her for days before she contrived to marry him if he agreed to wear spectacles before consummating their marriage. On their wedding night, at last, he put the eyeglasses on and got his first good look at the eighty-two-year-old bride. Outraged at first, he learned it was all a hoax and that she actually came to America to seek and designate him as her heir.

Poe usually leaves no doubt about the shortcomings of his narrators (in this case, shortsightedness). With humor and irony, he maintains a critical distance from them that makes his writings layered and ambiguous. When the narrator defends the veracity of love at first sight, he seems to remain oblivious to what has happened to him. The joke concerning the narrator's poor eyesight quickly sabotages his reliance on sight in matters of love. Yet in this tale, love is not necessarily blind either. In the opera, the narrator seems to betray a less-than-romantic interest when he admires the jewels of his newly discovered true love. Poe adds another ironic or oedipal twist by having the narrator fall in love with not just a mother figure but his actual great-great-grandmother.

Poe accentuates the ambiguity of these introductory lines by undermining the primacy of the sense of sight in love at first sight. The riveting glance only comes second to securing something beyond the eye's reach, referred to as "psychal fetters."[3] The expression is closely associated with what Poe called "psychal impressions," which are mentioned in "Mesmeric Revelation" (1844) and discussed in more detail in his 1846 marginalia. According to Poe, psychal impressions elude not just the sense of sight but all five senses. These types of impressions arise unexpectedly during a quasi-somnambulic state, "at those mere points of time where the confines of the waking world blend with those of the world of dreams." With his strange, ambiguous, and ironic tales, Poe claims that he yearns to produce these impressions even though they remain "beyond the compass of words."[4]

In "The Spectacles," electromagnetic interaction provides a striking new explanation for the formation of psychal fetters. The introductory passage quoted above appears riddled with worn-out analogies about the nature of love. Yet upon closer inspection, they are in fact highly unusual. During the first half of the nineteenth century, the norm was to rely solely on either magnetism *or* electricity to describe romantic attraction (i.e., as a magnetic attraction, *le coup de foudre*, etc.). Remarkably, "The Spectacles" suggests how magnetism and electricity together convey the workings of the invisible interactions behind human affections.

Due to "modern discoveries" concerning "magnetœsthetics" and "electric sympathy," Poe's narrator feels compelled to redefine the affective volatility of love as an electromagnetic interaction. This new take on love contrasts with previous ones because it depends on a physical model of relation marked by difference (fig. 7). What links electricity and magnetism bridges domains that are phenomenologically dissimilar. The two natural forces manifest distinct behaviors and obey physical laws that are not the same. Electromagnetism therefore manifests a metonymic power that connects through difference and, in "The Spectacles," brings a fresh perspective on human interaction.

Recent findings in "magnetœsthetics," or "ethical magnetism," have rendered this metonymic power more apparent. The expression *ethical magnetism* seems pompous and vague and participates in the display of the narrator's cluelessness. It also alludes, however, to the way magnetœsthetics comes with a set of structuring laws and values, imposed by the popular or current empirical understanding of (electro-)magnetism that shapes, through analogy, the comprehension of other phenomena. Since the pre-Socratics, these analogies have been an integral part of the Western ethos, particularly in its repeated attempts to probe and represent puzzling yet perennial ideas such as life, death, love, desire, polarity, and metonymic as well as metaphoric relations. When Poe coined the word *magnetœsthetics*, he gave this ethos a name that highlighted the historically omnipresent and changing (electro-)magnetic models that have shaped the arts and sciences and continue to impact all aspects of modern life.

In "The Spectacles," Poe performs an early and playful examination of a new ethos that emerged in the wake of electromagnetism's discovery. The story reinterprets the long-standing model of the magnetic chain. The association of magnetœsthetics with the image of psychal fetters dates back to Plato, who in *Ion* relied on a magnetic chain to convey the elusive nature of poetic enthusiasm and mimesis. This singular apparatus, made up of iron rings hanging from a magnet, has been a popular tool to represent and conceptualize various types of connections ranging from the human to the divine ever since.

When Poe created the word chain *magnetœsthetics*, he must have had in mind the recent publication of Percy Bysshe Shelley's "A Defence of Poetry" and his translation of Plato's *Ion*. Originally written in 1821 and appearing posthumously in an 1840 volume edited by his wife, Mary Shelley, these two texts have been very influential in literary studies and provide popular entry points into Platonic and Romantic views on aesthetics.[5]

In "A Defence of Poetry," Shelley is clearly a proponent of Hume's associationism. He considers the imagination as the main faculty for making fresh associations that "unveil the permanent analogy of things" and, in turn,

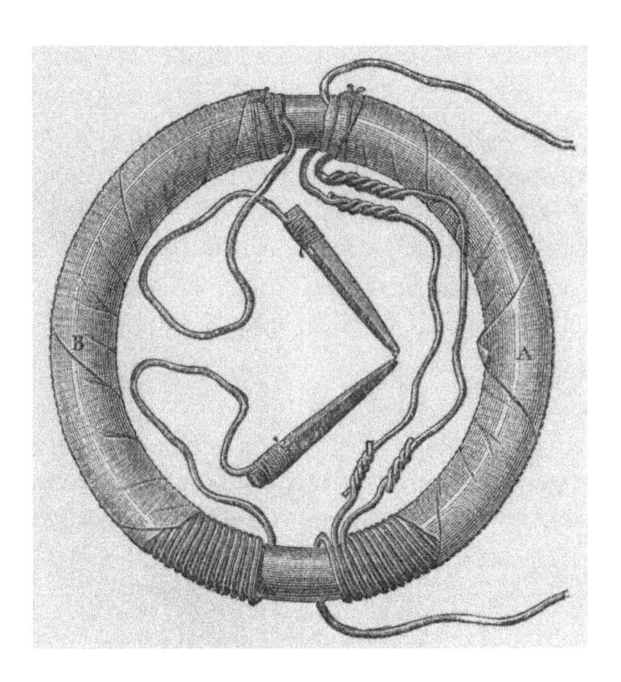

Fig. 7 Faraday conceived this "ring-coil apparatus" in August 1831. It played a crucial part in the experiments that led him to the discovery of electromagnetic induction. It consists of two separate coils wrapped around the opposite sides (B and A) of an iron ring. When an electric current starts to flow in one coil, it transforms the ring into a magnet, which in turn induces a current in the other coil. Faraday detected this induced current by the way it deflected a magnetized needle. This experiment provided an important clue that magnetism could induce electricity. It also showed that electromagnetic interaction produces a chain reaction. Friedrich Uppenborn, *History of the Transformer* (London: E. & F. N. Spon, 1889), 3.

generate new thoughts and knowledge about the world. To convey the workings and effects of poetic association, he principally relies on imagery inspired by the magnetic chain. For Shelley, poetry is like a magnet: "[It] enlarges the circumference of the imagination by replenishing it with thoughts of ever new delight, which have the power of attracting and assimilating to their own nature all other thoughts, and which form new intervals and interstices whose void for ever craves fresh food." Just as a magnet attracts and turns iron into another magnet, poetry is the binding power that brings people together and drives society's most significant achievements. Shelley adds that, despite the dark ages that have marked history, poetry has prevailed and constitutes "the co-operating thoughts of one great mind," which, "since the beginning of the world," have formed a long, continuous "chain": "The sacred links of that chain have never been entirely disjoined, which descending through the minds of many men is attached to those great minds, whence as from a magnet the invisible effluence is sent forth, which at once connects, animates, and sustains the life of all. It is the faculty which contains within itself the seeds at once of its own and social renovation."[6] A great magnetic chain of thoughts holds art, knowledge, and society together. Its binding power originates in the poet's imagination and, as Plato argues, susceptibility to the invisible influence of divine inspiration.

From Plato to Shelley, magnetic concatenation was a key empirical model to represent and conceptualize how different things could be related. Magnetization works through an invisible power that both connects and transforms, and it could therefore render more apparent—or even explain—other elusive and altering relations operating in the associative processes of the imagination, in human interconnection, and in history. In more general terms, the magnetic chain provides an effective and influential way to conceive of relation and difference within a vision of a world where continuity and discontinuity coexist. Depending on an author's inclination, this chain could emphasize either the metaphoric or the metonymic nature of a relation. Shelley foregrounds continuity based on magnetic similarity and spends little time on how this link between things is also established through the conspicuous difference implied by magnetic transformation.

At the end of "A Defence of Poetry," Shelley praises "the most celebrated writers of the present," who startle "with the electric life which burns within their words." Shelley's chain is electric as well as magnetic. Although he does not clarify how the two forces relate, Shelley's network of imagery suggests that they are metonymically linked. Such a link would refer to older theories concerning the unity of natural forces that were popular in the Romantic movement and whose affinity was usually attributed to a common origin. Since Shelley composed the essay the year following Oersted's discovery of electromagnetism, he might have been aware of electromagnetic interaction. In the case of Poe's "magnetœsthetics," the reference to electromagnetic interaction is conspicuous and, in turn, the display of metonymy more pronounced.

The symbolic and epistemological functions of the magnetic chain have traditionally overlapped with that of the more familiar model of "the great chain of being." The great chain of being remained a dominant vision of the order of the world well into the nineteenth century. It helped illustrate and spread hierarchical classifications based on a supposed gradual continuity among things. For its part, the magnetic chain came with its own set of characteristics, which led to specific interpretations of the great chain of being and especially the nature of its "links." Such magnetic interpretations have received little critical attention. For instance, Arthur Lovejoy's landmark intellectual history of the great chain of being traces how, from Plato to the Romantics, it expressed and problematized the principles of plenitude, continuity, and gradation. But he omits discussion of the chain's magnetic representations.[7]

"Seriality," a notion akin to the chain, emerged as the most common ordering pattern across disciplines during the nineteenth century. Nick Hopwood, Simon Schaffer, and Jim Secord have examined its scientific, cultural, and ideological legacies and how it posed "pervasive and prominent questions about continuity versus discontinuity" and "the play of difference [. . .], and the sequential display versus the array that could be seen at a glance."[8] Series emphasize various types of continuity that can guide the search for a "missing link" and can also provide a convenient arrangement for unrelated fragments, revealing a connecting pattern operating beyond their apparent discontinuity. In the nineteenth century, series helped visualize metonymic as well as metaphoric relations. Although the latter has dominated models associated with the great chain of being, as I showed in the introduction, from Plato and Kircher to Faraday and Maxwell, the former has remained an integral part of (electro-) magnetic concatenation and serialization.

During the second half of the nineteenth century, electrical experimenters helped highlight the scientific value of series based on discontinuous elements as they laid the foundation of particle physics.[9] Here I explore earlier types of series associated with the discovery of electromagnetism and the image of the chain that emerged during the first half of the nineteenth century. I claim that they anticipate this later shift toward discontinuity and metonymic relation. What Poe describes in the notions of magnetœsthetics and psychal fetters provides a remarkable example of such serial models, where discontinuity was not simply subsumed within a grand vision of continuity and metaphoric relations.

"The Spectacles" refers to "modern discoveries" in magnetœsthetics that I will examine through other relevant examples in Poe's oeuvre and the wider historical context informing them. These discoveries included changes in the physical understanding of magnetism and of the critical role it played as a binding power that made the great chain of being cohere. They also comprised the invention of a new medical practice known as animal magnetism, which helped bring forth fresh models of interconnection through the way it redefined the nature of the healer-patient relation and inspired many to rethink the link between life and death. Poe's ironic and fragmentary style manifests a metonymic shift that, I contend, amplifies a similar yet less familiar shift signaled by these "modern discoveries."

Poe's conception of electromagnetic psychal fetters offers a perspective that has been overshadowed by scholarship on the impact of Morse's telegraph on the nineteenth century. "The Spectacles" appeared when Morse was showcasing his revolutionary invention with a line connecting Baltimore and Washington, DC. The notion of psychal fetters brings to mind the lightning speed and intangible force employed by Morse's telegraph. As Adam Frank has shown, Poe makes references to the telegraph in his tales to provide a more material foundation for telepathic phenomena associated with animal magnetism, thought, and writing.[10]

Other insightful studies have focused on the cultural, epistemological, social, and political influences of the telegraph.[11] They tend to consider the telegraph as an "electrical" technology that changed the meaning of various types of human connection. What these works have glossed over, however, is the significance of the electromagnetic motor driving Morse's telegraph. When Morse found out about electromagnetic induction, he devised a mechanical stylus that moved according to the magnetic pull induced by an electric current, which could be sent over a long distance. Interruption in the flow of the electric current spaced out the marks on a paper roll. Morse also invented the famous code that translated these dots and dashes, printed by the electromagnetically powered stylus, into letters.

The interaction between electricity and magnetism that marks the opening lines of "The Spectacles" is a direct reference to the electromagnetic motor that made the telegraph possible. As a contemporary witness to the invention of the telegraph, Poe understood the new technology in more electromagnetic terms than later authors, who tended to ignore the symbolic potential of the metonymic relation between the two forces. By shifting the critical attention from telegraph lines to electromagnetic chains such as psychal fetters, what follows brings to the fore an older and influential network of metonymic analogies that informed early telegraphic imagery and that Poe associated with mesmerism.

A contemporary of Faraday, Poe was one of the first major literary figures to mobilize electromagnetism as a transformational motor to rethink human and cosmic interactions. Throughout his oeuvre, electromagnetic difference and relation turn his texts into Romantic machines dedicated to the production of fresh ideas, models to reorder the world, and, ultimately, psychal impressions. In "The Spectacles," the magnetic chains evoked by magnetœsthetics and psychal fetters get reimagined in electromagnetic terms that emphasize the metonymic complexity of affective experiences and the invisible links they form among people.

Such electromagnetic models had recently been invoked by influential proponents of animal magnetism, who used them to represent and explore the nature of human cognition, will, and interaction. Poe's interest in animal magnetism, and how it was a critical source of inspiration for both his fictional and his theoretical works, has been well documented.[12] What remains less clear is how electromagnetism influenced his understanding of animal magnetism.

Often referred to as mesmerism in honor of its inventor, Doctor Franz Anton Mesmer (1734–1815), animal magnetism played an integral part in the rise of modern psychotherapy, particularly due to its pioneering contributions to the domain of hypnosis and talking cures.[13] After initial therapeutic success turned sour in Vienna, Mesmer moved to Paris in 1778 to trumpet the wonders of his cures. Robert Darnton considers Mesmer one of the most important figures of the prerevolutionary era and places his name "somewhere near Turgot, Franklin, and Cagliostro in the pantheon of that age's most-talked-about-men."[14] Mesmer had such success in Paris that his fame quickly spread across Europe and to North America, converting countless physicians, writers, politicians, mystics, and charlatans to the doctrines and practices of animal magnetism.

Inspired by the invisible gravitational pull of planets and the attraction-repulsion of magnets, Mesmer claimed that organic bodies emanated a similar kind of animating fluid. A believer in the unity of natural forces, he first referred to these emanations as "animal gravity" before settling for "animal magnetism." Mesmer's analogy between mineral and animal magnetism prompted his decision to channel and control the body's magnetic fluid through the application of magnets on and around his patients' magnetic poles.

The magnet has always been—and continues to be—an integral part of medical science. In the West, the medical history of the magnet starts with Thales of Miletus (ca. 624–547 B.C.E.), who linked the magnet and the human soul based on the fact that both are endowed with the power of animation.[15] Hippocrates of Cos (ca. 460–360 B.C.E.) prescribed the lodestone to stop bleeding. Pliny the Elder (23–79 C.E.) recorded a treatment recommending pulverized lodestone for burns. Ever since Galen of Pergamum (129–199 C.E.), the attractive power of the lodestone has often been extended to draw out "bile," "phlegm," or corns. It was also believed to be a strong aphrodisiac. Magnetic attraction did prove an effective remedy against iron poisoning and provided a surgical tool to extract objects made out of iron, such as arrows, needles, or metallic particles, from the body. The development of smaller and more powerful magnets during the second half of the twentieth century facilitated the invention of practical vascular catheters, which can be teleguided by the manipulation of an artificial magnetic field immersing the patient. Today in the domain of medical diagnosis, Magnetic Resonance Imaging (MRI) is among the most advanced imaging technologies.

Magnets had also been used in the treatment of nervous diseases at least since the sixth century by Aetius of Amida. In the twelfth century, Hildegard of Bingen relied on lodestones to cure bad and malicious tempers, and by the sixteenth century, Paracelsus used their "repulsive pole" to treat epilepsy. More recently, it has been observed that electromagnetic fields can provoke physiological changes in the brain, and research is presently being conducted on their value in treating psychological disorders such as depression.[16]

The large number of parodies that Mesmer incited during his lifetime provides a clear measure of his fame. For instance, in Mozart's *Così fan tutte* (1789), a maid mockingly impersonates Mesmer as she wields the supposed healing power of a magnet above ersatz patients. Mesmer's handling of magnets stuck to his image, even though he quickly realized that he could achieve the same therapeutic effects by simply channeling his own "magnetic fluid" through his eyes and hands. He theorized about a pervasive and imponderable fluid akin to a magnetic effluvium to explain the strange influence he had on his patients.[17] According to Mesmer, illness resulted from an obstruction in the circulation of the body's magnetic fluid. He believed that he could project his own magnetic fluid to help reestablish its harmonious flow and, in turn, the patient's health. Such projection could provoke a "crisis" manifested by convulsions and swoons.

Mesmer's therapeutic method depended on holistic ideas concerning the unity of natural forces that had been popularized in medical practices since the vitalist doctrine of Paracelsus.[18] Mesmer's magnetic fluid theory shared a lot in common with that of lesser-known contemporary doctors who, for their part, used an "electric fluid" for similar treatments.[19] Mesmer's patients could gather around a large box called a "baquet"—a kind of reservoir for the magnetic fluid that contained iron filings and glass bottles organized in circular patterns. The design for this magnetic healing machine recalls that of eighteenth-century electrostatic devices like the Leyden jar.[20] Famous electrostatic experiments that consisted of transferring an electric shock across a chain of people holding hands must have also inspired Mesmer.[21] He describes these patients, connected to the baquet via iron rods and to their neighbors via the contact of their thumbs, as forming a magnetic chain of "one contiguous body."[22]

The rapid success of animal magnetism brought it under the scrutiny of the institutions of the medical establishment. They saw animal magnetism as a threat to their own (shaky) practices and in turn forcefully tried to discredit it through the publication of hostile reports. In retrospect, Mesmer's analogy between mineral and animal magnetism remained fruitful by stressing the importance of the imperceptible dynamic within the healer-patient relation. As the ancestor of hypnosis and suggestion, animal magnetism has recently been acknowledged by historians to have constituted a major step toward modern psychotherapies. They consider the "magnetic" influence Mesmer exerted on his patients as one of the first medical practices to reveal the central role the unconscious plays in the healing process.

Mesmer's loyal disciple, Amand Marc de Chastenet, Marquis de Puységur (1751–1825), now receives much of the credit for initiating what today are commonly called "talking cures."[23] In May 1784, Puységur discovered that with the help of "magnetic passes," some of his patients would fall into a state similar to sleep but would retain their ability to interact with the magnetizer. In this first case of artificially induced, or "magnetic," somnambulism, Puységur observed that his servant, Victor Race, was developing remarkable gifts.[24] Under the influence of the magnetizer, Victor's oral communication skills improved; he started to diagnose his own case, predict its course, and find its cure. After coming back to his wits, Victor did not remember what happened during his somnambulic state. Puységur's report on how, under the magnetic trance, Victor turned into the mouthpiece of another self is arguably the first credible medical record of the entity that would later be labeled the "unconscious."

Following the 1820 discovery of electromagnetism, studies in animal magnetism quickly began to invoke it as empirical evidence for the puzzling human interactions Mesmer and Puységur had brought to the fore. Electromagnetism was a striking confirmation of the unity of natural forces, making Mesmer's analogy of animal magnetism more plausible. It also lent support to those who perceived an intimate link between animal magnetism and other medical practices that based their ideas on Luigi Galvani's "animal electricity."

At the time Poe was writing "The Spectacles," mesmerists increasingly associated electromagnetism with alleged telekinetic phenomena produced by somnambulists or magnetic chains of people. As a transformational motor that did not simply obey the classic laws of mechanics, electromagnetically induced motion could help explain puzzling mesmeric interaction with objects. The craze for telekinetic phenomena culminated in 1853, when Faraday published two articles on "table turning." Across Europe and North America, the press reported that tables exhibited a mysterious motion in the presence of a group of people touching them and each other with their hands. Theories abounded as to what made tables "turn," "dance," "tip," "levitate," or "rap." Spiritualists believed that the dead produced these strange manifestations to communicate with the living. More scientifically inclined mesmerists and witnesses tended to attribute them to physical interaction akin to electromagnetism: the people around the table formed a chain of voltaic batteries, and their multiplied animal electricity induced a magnetic effect that made the table turn like a magnetic needle near a current-carrying wire.[25]

At one point, Faraday had given mesmerism some serious thought. In 1838 he observed the somnambulists of famed magnetizer John Elliotson, who was a professor of medicine at University College London. Although he quickly lost interest, his 1846 discovery of diamagnetism, which demonstrated that all matter (including living matter) is susceptible to magnetism, established his authority among supporters of animal magnetism seeking to establish its scientific foundation.[26]

Faraday did not get publicly involved with the solicitations of mesmerists until the widespread fascination provoked by table turning. In an article published in the *Medical Times and Gazette*, he announced that he had conducted a series of experiments to test the claims of table turners attributing the effect "to electricity, to magnetism, to attraction, to some unknown or hitherto unrecognized physical power able to affect inanimate bodies—to the revolution of the earth, and even to diabolic or supernatural agency."[27] Through the design of clever experimental devices, he refuted these claims and proved that table turning was the product of "a *quasi* involuntary muscular action" from the people touching the table. What the exact nature of such nonconscious movement was remained unanswered, though, and continued to fuel scientific and metaphysical speculations until the end of the nineteenth century.

Like countless other writers of the nineteenth century, Poe drew inspiration from the findings and popularity of animal magnetism.[28] His tales often rely on somnambulic states to convey strange mental and affective manifestations. In "The Spectacles," the seduction of the shortsighted narrator is complete when he hears his Dulcinea interpret the grand finale of Bellini's popular opera *La Sonnambula* (1831). The opera tells the story of a betrothed village girl who, unaware that she is a sleepwalker, provokes a scandal when she unwittingly goes into the room of a handsome lord. Poe's narrator ironically misses this musical reference to the nonconscious dimension of romantic attraction[29] when he confesses that her singing made him feel something beyond "electrical." Earlier, he had called this connection "a *magnetic* sympathy of soul for soul" (Poe's emphasis).[30]

"The Spectacles" appeared at the same time as other, better-known tales where mesmerism figured more prominently. "A Tale of the Ragged Mountains" (1844), "Mesmeric Revelation" (1844), and "The Facts in the Case of M. Valdemar" (1845) show that Poe was well versed in the latest theories of animal magnetism. These tales even incorporated scenes and theoretical speculations that he directly lifted from popular works on mesmerism such as Joseph Philippe François Deleuze's *Practical Instruction in Animal Magnetism* (1837, 1843) and Chauncy Hare Townshend's *Facts in Mesmerism, with Reasons for a Dispassionate Inquiry into It* (1840). The former is a translation from a famous French student of Puységur in which the author considers animal magnetism and electricity to be profoundly different yet possibly related and states that "recent discoveries of M. Œrsted, of M. Ampère [. . .] may give us some light upon [the magnetic fluid]."[31] The latter book made a strong impression on Poe, who deemed it "one of the most truly profound and philosophical works of the day—a work to be valued properly only in a day to come."[32] Townshend discusses at length the analogy of the mesmeric medium with other "imponderable agents" such as electricity, light, and heat. Referring at times to electromagnetism, he elaborates on a theory of the unity of natural forces where animal magnetism and electricity share an "affinity" despite their fundamental difference. The model he invokes to describe their relation is metonymic: "We are contending—not for the identity of the [mesmeric] agency with the electric, but for the propinquity."[33] Before the word *contiguity* came to the fore, Roman rhetoricians such as Cicero relied on the term "propinquity" (*propinquus*) to describe the relation of proximity that characterizes metonymy.[34]

In "The Spectacles," the electromagnetic nature of psychal fetters provides an important clue that Poe was fascinated not just by animal magnetism but by what I call "animal electromagnetism." Unlike Mesmer's animal magnetism, animal electromagnetism has incorporated the transformational motor unveiled by Oersted, Ampère, and Faraday into its theoretical framework. Whereas continuity dominated Mesmer's holistic theory of the mesmeric fluid, which suggested a direct connection between patient and healer, animal electromagnetism implied a clear distinction or discontinuity among the powers it brought together and, in turn, placed a greater emphasis on relations of contiguity. Concatenation in animal electromagnetism occurred through a binding power that was more clearly marked by metonymy.

Poe's most detailed version of a theory of the unity of natural forces appears in his last major work, *Eureka: A Prose Poem* (1848), which presents a detailed poetico-scientific conception of the cosmos where continuity and discontinuity could cohabit. Poe describes the universe in terms of a Romantic machine that endlessly expands and contracts like the "Heart Divine." In prose that mixes scientific knowledge with metaphysical and spiritual speculation, he writes about the attractive and repulsive forces that make this cosmic machine move. He claims that the innate tendency of atoms to reunite into a primordial "Unity" is rendered manifest by "the principle of Newton's gravity." On the other hand, the repulsive force that prevents atoms from fusing back together is like "that which we have been in the practice of designating now as heat, now as magnetism, now as electricity; displaying our ignorance of its awful character in the vacillation of the phraseology with which we endeavor to circumscribe it."[35] To simplify his task, and to provide empirical grounds for his speculation, Poe picks electricity to illustrate the force responsible for maintaining discontinuity. Since electricity relies on the proximity of differently charged bodies to generate a spark, it expresses not only the inherent movement toward the original state of homogeneity but also the repulsive force that resists this movement and preserves difference. As in "The Spectacles," Poe reiterates his conviction of the unity of electricity with other spiritual principles such as life, consciousness, thought, and mesmerism. He finally replaces the "equivocal terms" gravity and electricity with "the more definite expressions": attraction and repulsion, respectively. Repulsion thus became the latest metonymic substitution in a chain of equivocal terms that included electricity, magnetism, thought, and mesmerism and that Poe associated with the power of creating difference.

In *Eureka* and "The Spectacles," when Poe describes material, spiritual, and human interactions, he relies on animal electromagnetism and its mixture of mesmerism and electromagnetic interaction and transformation. The impact of animal electromagnetism on his oeuvre is most apparent in his mesmeric tales, where he mobilizes its metonymic power for his ironic reinterpretation of the function of death in the great chain of being. In what follows, I will show how Poe parodied another influential treatise on animal magnetism, Justinus Kerner's *The Seeress of Prevorst*, to undermine the continuity of the great chain of being through a shift from a metaphoric to a metonymic conception of the relation between life and death. This shift depended in great part on phenomena associated with magnetism that functioned as intermediaries among the spiritual, human, animal, vegetal, and mineral domains of nature. These intermediaries could connect domains considered fundamentally different because they appeared to partake in more than one. Their connective function depended then on the ambiguity of their status, which was necessary to make nature cling together but which also undermined the sovereignty of each domain.

Magnetic Animism and Death

"Mesmeric Revelation" (1844) appeared the same year as "The Spectacles." Although it is today one of Poe's lesser-known tales, it elicited considerable interest from contemporary readers and was Baudelaire's first translation of his work. The tale opens with a paragraph summing up what so many people found fascinating about animal magnetism at the time:

> WHATEVER doubt may still envelop the *rationale* of mesmerism, its startling *facts* are now almost universally admitted. Of these latter, those who doubt, are your mere doubters by profession—an unprofitable and disreputable tribe. There can be no more absolute waste of time than the attempt to *prove*, at the present day, that man, by mere exercise of will, can so impress his fellow, as to cast him into an abnormal condition, of which the phenomena resemble very closely those of *death*, or at least

> resemble them more nearly than they do the phenomena of any other normal condition within our cognizance; that, while in this state, the person so impressed employs only with effort, and then feebly, the external organs of sense, yet perceives, with keenly refined perception, and through channels supposed unknown, matters beyond the scope of the physical organs; that, moreover, his intellectual faculties are wonderfully exalted and invigorated; that his sympathies with the person so impressing him are profound; and, finally, that his susceptibility to the impression increases with its frequency, while, in the same proportion, the peculiar phenomena elicited are more extended and more *pronounced*.[36] (Poe's emphasis)

The narrator, "P.," calls magnetic somnambulism "sleep-waking," a term that Poe lifted from Townshend's *Facts in Mesmerism*. Predating mesmerism, the more familiar expression "sleepwalking" does not emphasize enough the state of heightened wakefulness—called at times clairvoyance, second sight, or lucidity—of magnetically induced somnambulism. The association of clairvoyance and magnetism participates in the tradition inaugurated by Plato's description of "enthusiasm" in *Ion*. At the beginning of the dialogue, Plato plays with the opposition between being asleep and awake as he describes the rhapsodist's magnetic enthusiasm in terms of the loss of the intellect accompanied by a heightened state of wakefulness. In "Mesmeric Revelation," willpower replaces divine power. Through acts of will, the magnetizer provokes the clairvoyant state.

Sleep-waking heightens perception "through channels supposed unknown," which reach "matters beyond the scope of the physical organs" and generate states that "resemble very closely those of *death*." This popular nineteenth-century analogy between death and mesmerism harks back to older beliefs that considered mineral magnetism linked to the mysteries of life and death. The earliest application of the compass in China was for the placement of buildings and tombs.[37] Since Thales and more recently Gilbert's proof of geomagnetism, a type of animism inspired by magnetism conveyed the occult influence between bodies (such as planets, people, or in spirit-matter dualism) in terms of a magnetic "soul" and its apparent ability to move or depart from a magnet. As discussed further in chapter 3, Gilbert's magnetic philosophy opposed the then-current scholastic conception of Earth as the motionless dead center of the universe. He argued the earth was alive and that it possessed its own magnetic soul, which was responsible for attracting the needle of the compass. Kircher, for his part, claimed that the power of attraction and repulsion of the lodestone manifested the inherent power of sympathy and antipathy structuring the cosmos, associating the former with life and the latter with death.[38]

In his classic study on the conceptual pitfalls that have historically misled scientific thinking, Gaston Bachelard discusses the problems posed by animism, using the example of the great chain of being and how mineral magnetism became a way to extend the chain's reach beyond the domain of life or the animate. His case study focuses on M. de Bruno, a magnet specialist from the end of the eighteenth century and "fine observer," who contended that

> the magnet offers us this particularity [nuance] of bringing living nature closer to nature that is inanimate; it reveals itself to us in the union of stone and metal, and in the latter this principle of life still spreads out with more energy. This amazing stone presents us with the wonders we marvel at in the fresh-water polyp, that extraordinary plant or rather animal which serves to link the vegetable to the animal kind [genre]. The magnet, like this polyp, can be cut in a parallel or transverse direction to its axis, and each new part becomes a magnet [. . .]. It is active nature that works in silence and invisibly.[39]

In the previous paragraph, de Bruno summarizes Comte de Buffon's influential conception of the great chain of being, where the human, animal, and vegetable "kind" cling together due to the connecting power of life.[40] De Bruno criticizes Buffon for omitting to connect the mineral kind to the great chain of being, intending to correct this oversight by using the lodestone as the missing link. For Bachelard, such an argument betrays a "real fetishism of life" that can lead to false hierarchies and assumptions.[41] Once limits and valorizations have been imposed on what should be considered "animate" or "inanimate," the need to keep nature a coherent whole gives rise to strange hybrid connectors or intermediaries that supposedly partake in more than one kind. De Bruno's attempt to integrate what appears inanimate into the domain of the animate describes a world where everything is to some degree endowed with life.

To defend his thesis, de Bruno compared the magnet to another famous intermediary, the freshwater polyp. In 1744 Abraham Trembley had undermined the distinction between the vegetal and animal domains in his study on various types of freshwater polyps. Considered a plant at the time, the polyp manifested animal behaviors such as autonomous movement. Yet the polyp looked like a plant too. Out of curiosity, Trembley proceeded to cut one polyp in two to see if it was capable of budding. A few days later, the separate pieces became two complete polyps. Unlike animals, the polyp could reproduce asexually. Trembley initially concluded that he had discovered a kind of "animal-plant."[42]

De Bruno saw a parallel between both the magnet's and the polyp's ambiguous positions in the great chain of being. He justified his principle of continuity between the living and the inanimate by pointing out that, similarly to the polyp, a magnet generates movement and, when sectioned, produces two new magnets. The essential features of the principle of life occur in the magnet's ability to move and reproduce. Yet magnetic reproduction operates like the morbid cut that makes two polyps out of one. Despite the monopoly of life or the supposed continuity of the inanimate with the animate, de Bruno's magnetic animism cannot do away with the discontinuity of a cut that marks a death and a birth in the reproductive chain of life.

The Seeress of Prevorst and Haunted Writing

The association of death and animal magnetism reached a peak of fascination with the 1829 publication of Justinus Kerner's *The Seeress of Prevorst* (translated into English in 1845). In this biographical work, Kerner retraces the life of Friederike Hauffe, an extraordinary magnetic somnambulist endowed with otherworldly powers and susceptible to "electricity, in all its forms."[43] Kerner knew Hauffe personally. He had been her magnetizer and confidant. *The Seeress of Prevorst* chronicles her short and tragic life and includes philosophical passages inspired by what she said and did during her magnetic states. Many of these interpretations revolve around the themes of death, mourning, and the great chain of being. Poe drew freely from this work for his mesmeric tales, knowing that it would resonate with his readership, but he added his own ironic spin—not only for humorous effect but to undermine Kerner's animist grip on the great chain of being. A closer look at Kerner's work clarifies the source of Poe's literary irony and how he wielded it to unleash a metonymic power associated with the discontinuity of life and death.

In the introduction of *The Seeress of Prevorst*, Kerner argues that the realization of the transience of life through the death of a loved one provides a privileged access to "inner life." Mourning provokes inner life and is most clearly expressed in phenomena linked to magnetic somnambulism, or "magnetic sleep-waking." Clairvoyance, premonition, possession, or, in short, magnetic life usually occur in ordinary people after they have suffered the trauma of a loss that undermines the grip of the outer world, the intellect, and the body on their existence. In magnetic life, the body becomes unnecessary, "as it were [. . .] dead," and gives way to a kind of spiritual supersensibility.[44]

Kerner opposed Hegel's view that magnetic somnambulism is a degenerate state rendered manifest by sick people and that it should therefore be considered below the intellectual achievement of the "Spirit."[45] For Kerner, magnetic life was a "wonder" that exalted and transcended the human condition. Magnetic somnambulists reincarnated ancient mystics and martyrs, who did not suffer from mental illnesses but rather enjoyed the rare ascendancy of magnetic life.

Similar to the way de Bruno found in the magnet the intermediary connecting the animate to the inanimate domains, Kerner used the "magnetic man" to link divine immortality to earthly mortality:

> However superficially we observe the course of nature, we cannot help remarking that she always advances by minute steps—that her progress is a chain, of which no link is wanting—and that she makes no abrupt transitions. Thus, in the stone we see the plant—in the plant, the animal—in the animal, man—and in man, the immortal spirit. And as the wings of the butterfly are folded in the caterpillar, so in man—especially in certain conditions—the wings of a higher Psyche are revealed, ready, after his short earthly life, to be unfolded; and, by the magnetic man, before whom time and space are unveiled, we learn that there is a super-terrestrial world. The magnetic man is an imperfect spirit. In the polypus, which is the link between the plants and the brute creation, we see both an imperfect animal and an imperfect plant; whilst fixed to the earth like a plant, it stretches its arms into the animal world, and thus bears witness to it. And, in like manner, we see the magnetic man, whilst yet in the body, and enchained to the earth, putting forth feelers into the world of spirits, and bearing witness to that also. Such a striving after, and upward flight into, the world of spirits, we observe in all magnetic subjects; but never yet in so great a degree as in [Friederike Hauffe]. We have seen [. . .] how this nerve-spirit—arrested, as it were, in the act of dying—became sensible of the spiritual properties of all things.[46]

Magnetism, metamorphosis, or the polyp—all appear to mark the overcoming of a limit that allows the great chain of being to cohere. Without these imperfect intermediaries, nature would fall apart; it would itself be imperfect, disconnected, disordered. The magnetic man is one of the go-betweens that make nature perfect by showing that its discontinuities are only apparent. Yet the magnetic man can only do so because, unlike nature, he is not a self-contained whole or plenitude; instead, he is caught between two domains and functions like a Derridean supplement. Like the polyp and magnet, his purpose is to show that nature is a perfect continuum, though he also functions as a stand-in for a missing link.[47]

The last rung of the natural ladder is reached when death can be superseded. For this ultimate step, the magnetic man, "arrested, as it were, in the act of dying," supplements nature by showing that a state between life and death, or between the "immortal spirit" and the mortal body, exists. For the final transition to immortal spirit, Kerner chooses to use death to strengthen the monopoly of life. Despite showing signs to the contrary, Kerner's animist resolution of the unresolvable discontinuity presented by death and by his supplement, the magnetic man, reinforces the great chain of being, which in turn keeps the old "natural" hierarchy and domains intact.

In *The Seeress of Prevorst*, Kerner retraces the life and revelations of Friederike Hauffe, a magnetic woman who partook of the world of spirits like no one else. Magnetic life runs in Hauffe's family. Her grandfather, with whom she was close during her childhood, was known to have premonitions. He was the first to notice that at a young age Hauffe showed signs of unusual sensibility when she would react to the nearby presence of metals or graves. Despite reporting that she grew up in a region known for the "nervous derangements" of its inhabitants, that at one point she saw a "specter," and that she was confined to her room for a "considerable" time due to a strange hypersensitivity of the eye, Kerner underlines Hauffe's relatively happy and normal youth. He also testifies that, contrary to some rumors, she neither formed nor suffered from emotional ties during her adolescence.[48]

However, at nineteen, after learning that her parents had contrived to marry her to a distant cousin, she fell into depression and stopped sleeping for five weeks. According to other biographers, two years earlier she had developed a strong attachment to Minister Tritschler, who was just about the same age as her grandfather and happened to be well versed in the art of mesmerism. He became her "spiritual father."[49] The day of her engagement ended up coinciding with the funeral of her beloved Minister Tritschler, with Kerner writing that "she followed the remains to the churchyard. However heavy her heart was before, at the grave she became light and cheerful. A wonderful inner-life was at once awakened in her; she became quite calm and could scarcely be induced to quit the grave. At length all tears ceased—she was serene, but, from this moment, indifferent to everything that happened in the world."[50]

For Kerner, this event marked the beginning of her "proper inner-life," and confirmed the affinity between death and spiritual rebirth, sorrow and joy. A few months later, in February 1822, she dreamed that Minister Tritschler was lying in her bed wrapped in a shroud. When her father and two physicians tried to take her away from him, she cried in her sleep that only this dead man could cure her. Her husband finally woke her up, and by the next morning "she was attacked by a fever, that continued for fourteen days with the greatest violence, and which was followed by seven years of magnetic life, interrupted only by short, and merely apparent, intervals."[51]

In 1823 she lost her firstborn child soon after birth. This event was followed by a "nervous" illness accompanied by the unfolding of the magnetic power she had already experienced during her childhood. Among other treatments, her family sought the help of a magnetizer. She apparently learned how to magnetize herself too. While in a state of magnetic sleep, she became aware of her own remedies and experienced premonitions and acts of clairvoyance. Part of the cure implemented by her first magnetizer consisted of an amulet, which contained, among other things, a small magnet. By 1826 she was living in a state of quasi-constant somnambulism when Kerner took over her treatment. In 1828 the death of her father caused the final shock, from which she would not recover. As she had herself predicted, she passed away the following year.[52]

Most of *The Seeress of Prevorst* is dedicated to transcribing and commenting on Hauffe's revelations on magnetic life and her visions of the spiritual world. Hauffe does not fit the traditional notion of the patient. She participated in her own diagnosis, probed the beyond, and dictated her findings to Kerner. Saïd Hammoud argues that *The Seeress of Prevorst* displays a new form of writing, since it is often impossible to distinguish between the voices of the magnetizer and the somnambulist.[53] This approach to writing marks the beginning of an uncharted form of medical objectivity that was prompted by the puzzling nature of Hauffe's state of dissociation, which called for the participation and introspection of the therapist during the healing process. Such methodology would eventually receive full attention with the invention of psychoanalysis. For its part, Kerner's book emulates the magnetic life of the somnambulist as it records dictations from a foreign body. Like in Nancy's reading of *Ion*, Kerner's prose is a magnetic chain of interpretations that manifests a metonymic logic of *partage*, or "sharing voices." Instead of the divine Muses, death and haunting phenomena associated with mourning are the primary sources of this *partage*. Hauffe herself, the "Seeress of Prevorst," is haunted by multiple voices.[54]

She reflects the process of haunted writing through her own incorporation of the magnetizer-somnambulist relationship, particularly when her deceased grandmother comes forth as her "protecting spirit" and inner magnetizer. Kerner writes, "We have above mentioned how it once appeared to her that she was magnetized by her protecting spirit [. . .]. This happened again here (in Weinsberg) [. . .]. After magnetizing her, the spirit bade her rise and write—which she did—and told her that the writing would remind her to teach her physician to magnetize her in that manner. Mrs. [Hauffe] begged the spirit to magnetize her always; but it answered—'Had I the power of doing so, you would soon take up your bed and walk!'"[55]

Like a magnetizer controlling a somnambulist through willpower, Hauffe's body is under the sway of the ghost of her grandmother. Other figures reminiscent of the magnetizer—like the charismatic grandfather and his substitute, Minister Tritschler, the "spiritual father" well versed in mesmerism—appear to have been incorporated by Hauffe as she declaims, sometimes in High German and sometimes in verse, the latest Romantic theories on the mysteries of magnetic life.

As Nicolas Abraham and Maria Torok note in their cryptanalysis of the cast of family members and therapists that made up the Wolfman's inner theater and would surface individually and intermittently by taking over his actions, "the first incorporation attracted others as a magnet draws up iron filings."[56] Since Hauffe's "proper inner-life," or haunting, began when Tritschler passed away, the initial incorporation that triggered the other ones may have taken place at that time. The manifestation of the "magnetic condition" during her childhood points to an older, secret incorporation—what Abraham and Torok call a "crypt"—that seems related to her grandfather and grandmother.[57] Such cryptic incorporation would have resulted from the inability to mourn a loss that was linked to the troublesome experience of a shared forbidden pleasure and that had to remain secret to keep the mental makeup of the mourner intact. This secret would, however, cause a severe melancholia, which in Hauffe's case turned particularly malign following her arranged marriage and Tritschler's death, and it consequently drew in other ghosts like a magnet.

On the receiving end, Kerner faithfully recorded the revelations of the ghosts haunting Hauffe's unconscious. Kerner most likely channeled his own ghosts in the process.[58] In 1797, when Kerner was eleven, the acute abdominal pain from which he had been suffering for two years, and which was attributed to "nervous" origin, was cured by a magnetizer. His father was also afflicted with similar pains and passed away two years later. During the previous yearlong convalescence of his father, Kerner had a vision of his "double" kneeling by his sickbed and holding his hand. In 1801, before having a magnetic dream that he interpreted as a sign that he had to study medicine, Kerner completed multiple apprenticeships, including one with a carpenter who made coffins. Later, in 1812, during a period marked by the loss of his mother and a few close friends, Kerner sought to rehabilitate death by considering it a form of spiritual rebirth similar to the metamorphosis of caterpillars and magnetic sleep. *The Seeress of Prevorst* was meant to substantiate this kind of dialectical resolution of death into immortal life, which could be interpreted as Kerner's symptomatic refusal to mourn or accept loss and, in turn, as a sign of his haunting.[59]

Poe's Ironic Chain of Contiguous Interpretations

The international success of *The Seeress of Prevorst* makes it a key source for understanding widely held beliefs concerning magnetic somnambulism. These beliefs emerge in Poe's mesmeric tales, where they help render the supernatural natural, particularly in moments when magnetic characters oscillate between life and death.[60] "Mesmeric Revelation" makes direct references to Kerner's book, with passages stating that magnetic somnambulism resembles death in a way that "it resembles the ultimate life" and that there "are two bodies—the rudimental and the complete; corresponding with the two conditions of the worm and the butterfly. What we call 'death,' is but the painful metamorphosis."[61]

In this tale, the narrator-magnetizer, named "P," follows the laudatory opening lines cited above by describing his last séance with Mr. Vankirk, a patient on the brink of death who had become clairvoyant under the influence of mesmeric treatment. Due to intriguing otherworldly sensations experienced during their séance, Mr. Vankirk has started to doubt his own doubt about the immortality of the soul and needs P to question him while he is in a state of magnetic sleep to seek out possible revelations about the nature of matter, mind, thought, and God. Significantly, during the dialogue that ensues between magnetizer and sleep-waker, Mr. Vankirk is referred to as "V," which visually evokes the split state affecting the somnambulist.

As the tale ends and V passes away, death is the final mesmeric revelation. The unusually rapid decay of V's body implies that he has been dead for some time, which prompts P to conclude, "Had the sleep-waker, indeed, during the latter portion of his discourse, been addressing me from out the region of the shadows?"[62] Magnetic sleep could then make a dead man speak, while also rendering this fantastic tale plausible for nineteenth-century readers. They would have been familiar with reports concerning similar magnetic experiences and most likely have read or heard about Kerner's dramatic description of Hauffe's final departure:

> On the 5th of August 1829, she became delirious, though she had still magnetic and lucid intervals. She was in a very pious state of mind, and requested them to sing hymns to her. She often called loudly for me, though I was absent at the time; and once, when she appeared dead, some one having uttered my name, she started into life again, and seemed unable to die—the magnetic relation between us being not yet broken. She was, indeed, susceptible to magnetic influences to the last; for, when she was already cold, and jaws stiff, her mother having made three passes over her face, she lifted her eyelids and moved her lips. At ten o'clock, her sister saw a tall bright form enter the chamber, and, at the same instant, the dying woman uttered a loud cry of joy; her spirit seemed then to be set free. After a short interval, her soul also departed, leaving behind it a totally irrecognizable husk—not a single trace of her former features remaining. During her life, her countenance was of that sort that is borrowed wholly from the spirit within; for which reason, though many attempts were made, no artist succeeded in transmitting her features to the canvass. It is, therefore, not surprising that, when the spirit had departed, the face should no longer be the same.
>
> [. . .]
>
> On the 7th, the post mortem examination took place, conducted by Dr. Off, of Löwenstein. The body was found wasted to a skeleton.[63]

Poe drew heavily from this passage of the 1845 translation of *The Seeress of Prevorst* for the striking conclusion of "The Facts in the Case of M. Valdemar," which was published the same year and in which he reimplemented the same narrative structure of "Mesmeric Revelation," though expanding on its gruesome ending: "As I rapidly made the mesmeric passes, amid ejaculations of 'dead! dead!' absolutely *bursting* from the tongue and not from the lips of the sufferer, his whole frame at once—within the space of a single minute, or even less—shrunk—crumbled—absolutely *rotted* away beneath my hands. Upon the bed, before that whole company, there lay a nearly liquid mass of loathsome—of detestable putridity."[64] "The Facts in the Case of M. Valdemar" exploited the mid-nineteenth-century fascination with the association of animal magnetism and death so well that, despite the grotesque exaggeration of this last paragraph, it became a hoax when respectable scientific periodicals started to publish it.[65] Kerner also wrote a well-crafted sensational ending that rendered the seemingly supernatural rapid decay of Hauffe's body more palatable by implying that it could be corroborated by several witnesses, including a doctor mentioned by name.

Kerner and Poe have in common an obsession with death. But their approach to this elusive subject is completely different: Kerner, the true believer, wants to convince the reader of the monopoly of life with his earnest account of death as "the ultimate life," whereas Poe, the ironist, exploits the popularity of such reports to write a story designed not only to convince Kerner's gullible readers but also to unsettle them through a parodic appropriation that undermines the promise of a dialectical resolution of death into life, thereby reintroducing a fragmenting difference in the great chain of being.[66]

Gary Richard Thompson has shown how Poe leans on a conception of irony inspired by German Romanticism that enables him to approach his obsessions with death, loss, and meaninglessness from a critical distance.[67] In "The Facts in the Case of M. Valdemar," Poe tackles the themes of life and death but manages to keep an ironic and comic detachment (the two are closely related) through the ambiguity at the root of a successful hoax. The duplicity of such stories offers what Thompson calls a "multiple vision," which in Poe's oeuvre takes the form of a chain of paradoxical yet closely linked interpretations. This chain connects the credulous and incredulous and the supernatural and natural readings of the tale, prompting both laughter and perplexity.

Poe cannot then be taken seriously. Yet it is precisely at this moment, when he laughs at Kerner's expense, that he destabilizes the animist monopoly of life through an ambiguous tale that reemphasizes the profound—and consequently grave—discontinuity of life and death. Both states share an intimate relation not just marked by resemblance, as Kerner would argue. Poe appropriates *The Seeress of Prevorst* to construct a magnetic chain of interpretations that, through the multiple vision of irony, recasts the great chain of being in more metonymic terms. In this more complex chain, metaphoric links cannot simply absorb connections based on contiguity.

Poe found a metonymic power in animal electromagnetism linked to death and mourning that gives shape not only to his mesmeric tales but also to his most familiar legacy: modern detective fiction. In "Mesmeric Revelation," sleep-waking allows V, much like Hauffe, to access a higher perception that would have otherwise remained inaccessible. Although his domain is that of the metaphysical, the description of V's extraordinary power of detection echoes that of C. Auguste Dupin: "I cannot better explain my meaning than by the hypothesis that the mesmeric exaltation enables me to perceive a train of ratiocination which, in my abnormal existence, convinces, but which, in full accordance with the mesmeric phenomena, does not extend, except through its *effect*, into my normal condition. In sleep-waking, the reasoning and its conclusion—the cause and its effect—are present together. In my natural state, the cause vanishing, the effect only, and perhaps only partially, remains."[68] While magnetized, V perceives a nontrivial relation between cause and effect. What remains fragmentary and puzzling in a normal state becomes "a train of ratiocination" in sleep-waking. In private letters, Poe refers to a group of his literary works as "tales of ratiocination," especially those featuring Dupin's famed method of detection.[69]

Dupin initially appears in the tale many consider to be the first example of modern detective fiction, "The Murders in the Rue Morgue" (1841). Its narrator, who is Dupin's friend, notes that he achieves his most remarkable ratiocination in an altered state akin to sleep-waking: "His manner at these moments was frigid and abstract; his eyes were vacant in expression; while his voice, usually a rich tenor, rose into a treble which would have sounded petulantly but for the deliberateness and entire distinctness of the enunciation. Observing him in these moods, I often dwelt meditatively upon the old philosophy of the Bi-Part Soul, and amused myself with the fancy of a double Dupin—the creative and the resolvent."[70] Dupin's power of detection becomes especially acute in a volatile state where the creativity of the imagination works in tandem with the analytical power of reason.[71] At the end of the tale, Dupin explains that he was able to solve the mystery that eluded French policemen because, he boasts, they are only "head" while he is both "head" and "body." The year before "double Dupin" appeared, Townshend's *Facts in Mesmerism* underlined the dual nature responsible for the mesmeric, heightened sense of perception by calling magnetic somnambulism "sleep-waking." In "Mesmeric Revelation" and "The

Facts in the Case of M. Valdemar," Poe makes this duality even more apparent by inserting a dash between "sleep" and "waking." This visually accentuates the relation of contiguity that makes this chain of paradoxical mental states cling together. Dupin owes his special detective skill to a remarkable metonymic power that allows him to connect conscious and nonconscious cognitive faculties represented by the opposition between "head" and "body."

Like V in "Mesmeric Revelation," this metonymic power helps Dupin generate "a train of ratiocination." It takes the form of a chain that appears at first completely fragmentary to both narrator and reader until Dupin unveils its intricate relations. Before turning to the murder case, the narrator offers a preliminary example to convey what he calls Dupin's "method." The narrator recalls how they were once silently walking the streets of Paris when Dupin suddenly replied to him as if he had heard what the narrator had been thinking about for the past fifteen minutes. To explain how he was able to read his friend's mind, Dupin begins by enumerating a series of seemingly unrelated events that marked the narrator's behavior during their perambulation:

> The larger links of the chain run thus—Chantilly, Orion, Dr. Nichol, Epicurus, Stereotomy, the street stones, the fruiterer.[72]

Dupin then proceeds to unveil their relation by explaining how they shaped, through associations, what the narrator was thinking and how, by inferring those associations, he figured out his friend's train of thoughts. To read the minds of other people, Dupin's method depends on a Humean principle of associations and, more specifically, on an extraordinary ability to perceive metonymic links among what appears to be a series of random elements.

This method is part of the formula that has made detective fiction so successful and addictive. The reader is presented with a wide array of disconnected clues and is invited to solve the puzzle before the fictional detective closes the case. At the beginning of "The Spectacles," Poe, who claimed that all his tales shared a certain unity of composition,[73] also invites readers to detect a metonymic chain in the seemingly insignificant details shared by the narrator about his French ancestors. The narrator notes an interesting coincidence in the way the series of names constituting his lineage sound the same:

> Here [. . .] are *Moissart*, *Voissart*, *Croissart*, and *Froissart*, all in the direct line of descent.[74]

While the narrator identifies a chain marked by a resemblance deriving from the same suffixes, the reader can detect another type of link based on the contiguity of the French prefixes, which translate into English as "Me" (*Moi*)—"See" (*Vois*)—"Believe" (*Crois*)—"Cold" (*Froi[d]*). This metonymic chain tersely foretells the misadventure of the shortsighted narrator: "me" will "see" and fall in love and "believe" in this romance until the "cold" reaction of realizing it was a hoax organized by the French great-great-grandmother. "The Spectacles" invites the reader to become a detective and figure out the train of ratiocination underlying this chain of names. This less apparent chain announces that the romance will remain within the family and that the narrator is a kind of sleep-waker who speaks in multiple voices. It also reminds the reader that an effective method of detection involves metonymic as well as metaphoric chains of interpretations.

When Dupin applies his method to the mysterious murders of the Rue Morgue that have baffled French police, he reconstructs a chain of events out of disparate clues through inferences that he characterizes as a kind of induction: "You will say that I was puzzled; but, if you think so, you must have misunderstood *the nature of the inductions*. [. . .] There was *no flaw in any link of the chain*. I had traced the secret to its ultimate result [. . .]" (my emphasis).[75] In his seminal study on detective fiction and science, Régis Messac contends that inductive reasoning is where the two domains most clearly intersect.[76] Championed by Francis Bacon and central to the empirical sciences, the method of induction collects, organizes, and compares individual observations to infer a general conclusion.[77] Dupin practices it in the same way scientists specializing in historical investigation do. Similarly to a geologist or paleontologist, the detective compiles clues to reconstruct past events. He presupposes little or nothing in the process in order to keep all possibilities open. In "The Murders in the Rue Morgue," Dupin's inductive method allows him to identify the most unlikely of criminals: an orangutan.

Building upon Messac's study, Charles J. Rzepka calls this scientific approach "metonymic induction." Clues such as fingerprints or a tuft of hair left at the crime scene function like metonyms. They are fragmentary parts of a whole that the detective must piece back together to figure out what happened. The French police practice a kind of inductive method too. Yet Dupin is more effective because, as he reminds the narrator, his inductive reasoning depends on a combination of "head" and "body." This combination allows Dupin, through a kind of informed intuition, to make bold guesses and creative leaps that reveal the importance of details and metonymic links overlooked by the police.

The police's failure to solve the crime participates in Poe's wider critique of the scientific method. This critique culminated in *Eureka*, where, in the opening pages, the supposed author of a letter from a distant future attacks Francis Bacon and his followers for supporting a conception of inductive reasoning that ignores the crucial role played by "intuition" and the imagination in the process of discovery.[78] Poe appropriated the term *induction* for his own purposes, making it more all-encompassing but also looser and more difficult to define.[79]

Literary critics and philosophers of science have argued that Dupin's inductive reasoning anticipated what Charles Sanders Peirce (1839–1914) described as "abduction."[80] Peirce contended that, before inductive generalization, a crucial preliminary step occurs when new ideas or hypotheses deriving from puzzling observations emerge in the mind. He called this step abduction and, particularly in his later works, characterized its mode of operation in terms of intuition, guesswork, and imaginative and nonconscious faculties. Although the exploration of Dupin's method through the lens of Peirce's abduction has helped illuminate Poe's tales and their legacy in the development of modern detective fiction, abduction loses the contemporary resonance of the term *induction*, which Poe exploited to rethink its traditional Baconian meaning.[81] Mesmerism highlighted volatile and nonconscious cognitive states that, like many others, Poe compared to electromagnetic phenomena such as induction.

Before examining the wider impact of electromagnetic induction on the history of scientific reasoning, it is important to reiterate how its mesmeric appropriation contributed to rendering manifest relations of contiguity at work in cognition. In his tales of sleep-waking and ratiocination, Poe refers to such relations to describe how metonymic induction works. This type of inductive reasoning derives its power of detection from an interaction between head and body that underpins the way Dupin can read the mind of others. Rzepka characterizes this empathetic skill as "metaphorical rather than metonymical thinking."[82] Dupin identifies with others because he thinks *like* them. Yet what also makes this process of identification possible comes from "double Dupin," who embodies a relation of contiguity rendered manifest in sleep-waking when magnetic somnambulists become other.

As Kerner made clear in *The Seeress of Prevorst*, this metonymic process of identification is intimately linked with death and mourning. Andrea Goulet has recently underscored that one of the main legacies of "The Murders of the Rue Morgue" is the "haunting origin-point of violent death" that undermines the myth of detached scientific rationality and that has been attached to detective fiction ever since.[83] During the first half of the nineteenth century, empirical scientists mobilized the power of inductive reasoning to reveal that the present was marked by violent events from the past. They combed the globe for geological clues, from which they reconstructed past cataclysms and destructions. Their findings brought to the fore catastrophic events like extinction and dramatic environmental change that questioned the linear continuity of natural history and, by extension, the traditional framework that held together the great chain of being.

As human supplements linking the domains of life and death, sleep-wakers experienced haunting phenomena that also appeared to undermine this framework. Hauffe was under the sway of past tragedies involving death and family trauma that made her speak in multiple voices and enter higher states of perception. In "Mesmeric Revelation," V's train of ratiocination engages with metaphysical mysteries that appear only accessible from "the region of the shadows." Unsolved violent deaths transport "double Dupin" through time and space as they unleash his power of induction to recreate a complex chain of events.

Through irony and tales of ratiocination, Poe obsessively writes about death while inviting his readers to generate their own trains of ratiocination via inductive reasonings akin to those practiced both by empirical scientists and by sleep-wakers. By presenting fragmentary chains of puzzling observations, his tales are inductive apparatuses. They call upon the reader's imagination to make nontrivial associations that stress the importance of chains established through metonymic connections to unveil the truth.[84] "The Spectacles" provides important clues showing that Poe largely modeled the nature and function of these literary inductive apparatuses on newly found electromagnetic and mesmeric interactions. These interactions inspired influential mesmerists such as Townshend to redefine Mesmer's magnetic fluid theory in terms of the physical phenomena uncovered by Oersted, Ampère, and Faraday, giving birth to an influential yet less familiar branch of animal magnetism—animal electromagnetism.

One of Poe's famous contemporaries, Honoré de Balzac, also drew inspiration from animal electromagnetism as a metonymic model for inductive reasoning. In fact, Balzac pioneered the literary representation and exploration of animal electromagnetism. He did so in a less ironic style, which makes the mode of operation of his own literary inductive apparatuses more transparent and, in turn, helpful in clarifying the early appropriations and interpretations of electromagnetic contiguity.

2.

Induction Apparatuses

Electromagnetic Induction and the Volatility of Discovery

Honoré de Balzac pioneered the literary use of electromagnetic induction as a transformational motor and Romantic machine. Soon after its discovery in 1831, induction appears in his philosophical novel *Louis Lambert* in a striking new description of the volatile dimension of scientific thinking. What makes this analogy particularly remarkable is that it replaced a previous one inspired by Newtonian physics. *Louis Lambert* is a semiautobiographical story where the eponymous character is a genius who experiences a revelation about the secret nature of the universe when he realizes that he possesses the gift of clairvoyance. For the narrator of the 1832 edition of the novel, this sudden experience happened "as the fall of a pear became the primary cause of Newton's discoveries."[1] In the 1833 edition, it occurred "as the electric sensation always felt by Mesmer at the approach of a particular servant was the starting-point of his discoveries in magnetism."[2] What motivated Balzac to replace a Newtonian analogy for an unprecedented electromagnetic one?

Like Poe, Balzac developed a strong interest in animal magnetism. He was a firm believer in the powers of magnetizers and somnambulists and devoted many pages of *La comédie humaine* to promoting, documenting, and staging Mesmer's doctrines. Yet the change he made to *Louis Lambert* expresses more than just his interest in animal magnetism: it shows a remarkable knowledge of the latest research in electromagnetic science that Mesmer could not have known. In the new version of the novel, the interaction responsible for Mesmer's discovery of animal magnetism works like the lab apparatus that Faraday devised to unveil the phenomenon of induction (fig. 8). The approaching servant recalls a moving magnet that induces electricity in a nearby conductor represented by Mesmer's body.

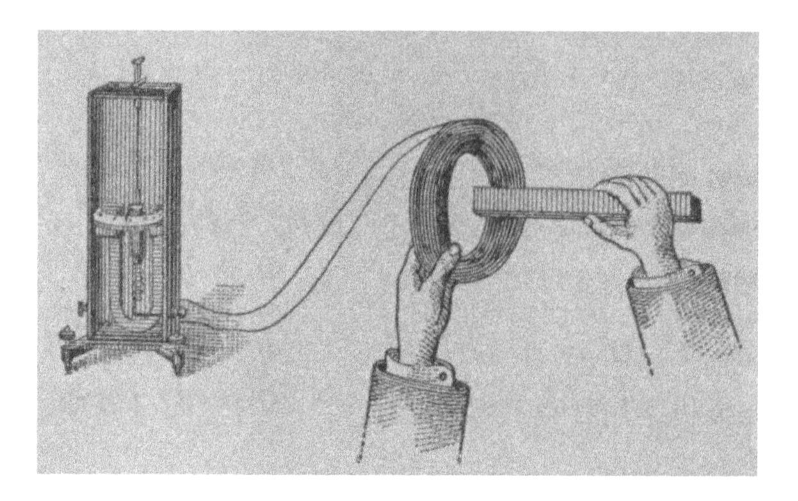

Fig. 8 Faraday explored the effect of induced currents with various laboratory apparatuses. This diagram represents one of the simpler ones, where the motion of a bar magnet induces an electric current in a coil of conducting wire. The magnetized needle of a galvanometer is used to detect the current. Henry S. Carhart and Horatio N. Chute, *Practical Physics* (Boston: Allyn and Bacon, 1920), 401, figure 426. Library of Congress, Washington, DC.

The influence of Faraday's electromagnetic discoveries on nineteenth-century literature has been the subject of insightful critical studies that have concentrated on the impact of his early formulation of field theory on the ideas of Ralph Waldo Emerson (1803–1882). Eric Wilson has traced how Faraday's empirical work on the identity of electricity and magnetism helped Emerson conceptualize the interconnectedness of the universe on firmer empirical grounds. He shows how Emerson's understanding of Faraday's speculations on the dynamic and immaterial fabric of the universe informed his unitary vision of mind, matter, language, and the sublime as variations of "spherules of force."[3] Sean Ross Meehan has noted that the notion of "spherules of force" provided "a crucial analogy for a material thinking of metamorphosis" that, in works published after 1850, lent support to a shift that Emerson operated from a metaphoric to a metonymic vision of the relations linking mind, matter, and poetry.[4] Laura Dassow Walls has also focused on Emerson's protofield theory, showing how it deeply informs his later philosophical and political thoughts and claiming that by following "Faraday's imagination farther than anyone else in his lifetime [. . .] Emerson rode the wave that led straight to Einstein."[5]

I contend that Balzac was the first major literary figure to jump in the same Faradayan "wave" that guided Einstein when he discovered the theory of relativity. Walls's image of "the wave" implies a cross-disciplinary line of thought sweeping across the nineteenth century, linking Emerson to Einstein based on the former's elaboration of a protofield theory. The example of Balzac demonstrates an older yet more direct foundation for this line of thought that originated in Faraday's induction apparatus. This device revealed the phenomenon of electromagnetic induction and the set of conceptual challenges that prompted Faraday to develop field theory. In *Louis Lambert*, Balzac refers directly to this induction apparatus because it manifested an alternative to spatial interaction sanctioned by Newtonian physics. In his seminal paper "On the Electrodynamics of Moving Bodies" (1905), Einstein invokes the same induction apparatus to anchor his elaboration of the special theory of relativity and its devastating critique of the Newtonian notions of absolute space and time. In both Balzac's analogy and Einstein's theoretical physics, Faraday's induction apparatus manifests a relation of contiguity between electricity and magnetism that became a critical tool for rethinking the nature of relation and difference in their respective fields. By shifting the critical attention from field theory to Faraday's induction apparatus, I reassert the central role this transformational motor played in the emergence of electromagnetic thinking.

Balzac's analogy is also remarkable because it signals the emergence of electromagnetic thinking by bringing together Faraday's induction apparatus and the scientific reasoning that bears the same name. When Newton observes the fall of a fruit and infers the theory of universal gravitation, he is an empirical scientist who practices inductive reasoning. In the revised version of *Louis Lambert*, Mesmer induces the existence of animal electromagnetism by identifying a connection between an unusual bodily experience and the presence of his servant. Although based on different physical models, these anecdotal accounts of scientific discovery aim to represent the same thing: the volatile aspect of inductive reasoning. Besides making observations, there is an unpredictable moment of insight when the scientist finds a previously unknown link among them that reveals a general principle.

In Mesmer's case, Balzac represents the volatility of this moment through a striking rapprochement of electromagnetic induction and inductive reasoning. Electricity and magnetism had been popular models to represent all kinds of volatile experiences. Yet the discovery of their mutual induction added another dimension to their volatility. As one force induces the other, they express the passage of a limit or marked difference that separates the two domains and that helps Balzac describe *another* threshold that inductive reasoning must cross to identify a new principle.

Balzac's electromagnetic analogy becomes especially significant considering later and more influential examples that also refer to both types of induction in the context of scientific thinking and discovery. In the 1840s, Poe's tales of ratiocination were inductive apparatuses that prompted the reader's imagination to solve mysteries through metonymic inductions akin to those of detectives and empirical scientists. As we will see, Einstein claims that he induced the theory of relativity based on observations concerning the puzzling relation of contiguity unveiled by Faraday's induction apparatus. In Einstein's words, "The phenomenon of the electromagnetic induction forced me to postulate the (special) relativity principle."[6] As mentioned in the introduction, Valéry invokes the induction apparatus of the dynamo to highlight a fundamental difference between mechanical and electromagnetic thinking and, in turn, the inferences that characterize their respective modes of detection. The metonymic link responsible for the mutual induction of electricity and magnetism presented a physical problem that pushed scientific thinking beyond the framework of Newtonian physics and, in the same movement (and as Balzac's analogy points out), also pushed it beyond the mechanist interpretation of inductive reasoning.

Balzac's reference to Faraday's induction apparatus thus offers critical insight into the emergence of electromagnetic thinking that, before addressing its legacy in the works of Einstein, needs to be resituated within the context of its origin—namely, the literary realism of *La comédie humaine*. Like Poe, Balzac was concerned with what moved things. In the avant-propos to *La comédie humaine*, he argues that his encyclopedic literary enterprise sought not only to depict postrevolutionary French society but to unveil the source of its "movement" and "social motor."[7] For Balzac, willpower was the main principle behind social and historical change. Throughout his fictional works, he develops a dynamic theory of the will by representing its ebb and flow, concentration, dissipation, and transformations. John Tresch considers Balzac an exemplary "mechanical Romantic" and has shown that electrochemical conversions and animal metamorphoses were crucial sources of inspiration for his conceptual framework.[8] In his meticulous study of *La peau de chagrin*, Sydney Lévy demonstrates how the steam engine provided Balzac with a concrete transformational motor to explore and convey the elusive nature of what set people and society in motion.[9] I examine the other main motors giving form and content to *La comédie humaine*: animal magnetism and electromagnetism.

Balzac's literary innovation is particularly indebted to magnetic somnambulists like Friederike Hauffe. Under magnetic sleep, they reportedly reached higher levels of oral expression, and their clairvoyance allowed them to bypass the traditional constraints of space and time and to dictate detailed descriptions of distant places—much like the omniscient narrator of Balzac's so-called realist novels. Building upon what Göran Blix has aptly called Balzac's "magnetic realism," I will show that an electromagnetic realism began to appear sporadically in his novels of the 1830s. In these particularly philosophical works, Balzac reinterprets Mesmer's magnetic fluid theory through the lens of newly found electromagnetic interactions. Balzac was particularly attuned to the implications of Faraday's discovery because he believed in the same Romantic idea that had guided Oersted to his discovery of a connection between electricity and magnetism—namely, the unity of forces.[10] Balzac was also a friend of André-Marie Ampère's son, Jean-Jacques Ampère. Jean-Jacques Ampère had an illustrious career as a literature professor and reportedly joked that his two greatest achievements came down to having met Balzac when he was both unknown and skinny.[11]

After tracing the function of electromagnetism in *La comédie humaine*, I will situate Balzac's rapprochement of Faraday's induction apparatus and inductive reasoning within contemporary debates concerning the meaning of *induction* in the history and philosophy of science. Faraday picked the term because of its currency in electrical science where, as opposed to *conduction*, it describes an influence that takes place through space without apparent contact. Although it implied a kind of action at a distance, the term remained vague enough to allow Faraday to circumvent the prevalent Newtonian theoretical framework of his time and, in turn, practice a more open-ended method of exploration and reasoning that formed the root of his innovations.

During the 1830s and 1840s, inductive reasoning was the subject of important controversies: most notably, the one between John Stuart Mill and a friend and collaborator of Faraday's, William Whewell. Although Whewell elaborated detailed and intricate theories to rationalize the nature of inductive reasoning, he conceded that "the process of induction includes a mysterious step, by which we pass from particulars to generals, of which step the reason always seems to be inadequately rendered by any other words which we can use; and this step to most minds is not demonstrative."[12] Mill, on the other hand, considered only what could be empirically demonstrated, and he contended that his inductive method did not involve such "a mysterious step," suggesting instead that it was a completely detached and algorithmic process.

According to Laura J. Snyder, what Whewell and Mill had in common was a nonsyllogistic view of inductive reasoning. Unlike the syllogistic logic of deduction, induction revealed a new principle by leading from the known to the unknown, which involved crossing a threshold between the same and the different. A metonymic imperative informed the practice of what they considered true inductive reasoning.[13] Considering these various usages of the term *induction* and Maxwell's praise of Faraday's indirect method of investigation, I argue that Balzac's image of the discovery of animal electromagnetism brings to the fore "a mysterious step" that undermines the myth of the detached scientist by including a kind of metonymic reasoning linked to the body in the process of discovery.

Magnetic Realism

Vautrin, the great villain of *La comédie humaine*, ultimately becomes the chief of police and exerts his remarkable power of detection for the opposing side. This forefather of other famous characters such as Dupin and Svengali initially appears in *Le père Goriot* (1835), where he is a kind of magnetizer who can throw "a cold and fascinating glance; men gifted with this magnetic power can quell furious lunatics in a madhouse."[14] At his most cunning moments, Vautrin is also Balzac's mouthpiece. Balzac presents his critical dissection of modern France as Vautrin reads the minds of other characters and reveals society's dark secrets and hidden motors. In *Le père Goriot*, when the narrator says that Vautrin "read the minds [*lut dans l'âme*] of the two young people," it implies the ability to see through the veneer of social behavior as well as through the physical barriers that protect the privacy of one's thoughts. Vautrin reads Rastignac's thoughts like an open book because "his eyes [. . .] seemed to go to the very bottom of all questions, to read all natures, all feelings and thoughts. [. . .] He knew or guessed the concerns of everyone about him; but none of them had been able to penetrate his thoughts."[15] Unlike Dupin's mind reading, which seems to come from his exceptional analytical skills and ability to identify with others, Vautrin's nonintrusive (in the medical sense) magnetic gaze overcomes physical barriers and challenges conventional definitions of space.

Such telepathic phenomena deeply intrigued Balzac. For instance, in *Le réquisitionnaire* (*The Conscript*, 1831), he explores the nature of "sympathies." This short philosophical novel portrays Madame de Dey's agony as she waits through the night for the return of her beloved son, who had been captured during a military operation and who had made plans to escape that same night. Her son does not come back, and without a clear reason, Madame de Dey suddenly dies the following morning. This tragic ending prompts the narrator to conclude with the following thoughts: "The death of the countess had a far more solemn cause; it resulted, no doubt, from an awful vision. At the exact hour when Madame de Dey died at Carentan, her son was shot in the Morbihan. That tragic fact may be added to many recorded observations on *sympathies that are known to ignore the laws of space*: records which men of solitude are collecting with far-seeing curiosity, and which will some day serve as the basis of a new science for which, up to the present time, a man of genius has been lacking"[16] (my emphasis). With such stories, and many other examples throughout *La comédie humaine*, Balzac leaves no doubt that he counted himself as one of those "men of solitude" who sought to establish "a new science." His numerous and wide-ranging novels worked like an archive where he collected and organized various observations in an effort to induce the general principle underpinning puzzling phenomena such as "sympathies."

Dahlia Porter has shown that the rise to prominence of the inductive method during the seventeenth and eighteenth centuries reshaped not only the sciences but also literature and philosophy. Bacon's attempt to modernize and codify the inductive method led to challenging questions concerning how data should be compiled and how the step from individual observations to a general principle worked. Attempts to address these questions or to solve thoroughly "the problem of induction" informed the explosion of encyclopedic projects during the Enlightenment. Faced with the rapidly expanding production of information and knowledge, scientists, philosophers, and writers alike experimented with various systems and textual forms that could help them induce the hidden unity of disparate materials.[17] *La comédie humaine* is an encyclopedic account of French society that participates in this transdisciplinary tradition. Balzac claimed that his monumental oeuvre followed a "unity of composition," which, similar to the work of naturalist Étienne Geoffroy Saint-Hilaire, provided the organizational principle and formal coherence that turned his wide-ranging novels into a whole.[18] As announced in the aforementioned passage from *Le réquisitionnaire*, this unity of composition also rendered manifest the foundation of "a new science" that sought to unveil the nature of phenomena that "ignore the laws of space."

Balzac's conception of sympathies built upon his spiritual beliefs (Catholicism, Swedenborgism, Martinism) and materialist theories of cognition. In the 1842 avant-propos to *La comédie humaine*, he writes,

> In some portions of my long work I have attempted to bring to popular notice the amazing facts, or, I might say, *the marvels of electricity*, that exerts so incalculable an effect upon man; but how should those cerebral and nervous phenomena, proving the existence of a new moral world, disturb the sure and certain relations between God and his worlds? How should the Catholic dogmas be shaken by them? *If, by indisputable fact, thought is one day classed among those fluids which are known only through their effects*, and the substance of which evades our senses, however intensified by mechanical contrivances, it will be the same thing as when Christopher Columbus observed that the earth was round, or when Galileo demonstrated that it moved upon its axis. Our future cannot be affected by it. *Animal magnetism and its miracles*, with which I have been familiar since 1820; the interesting researches of Gall, the successor of Lavater; and all those who, for the last fifty years, have been studying thought as the opticians have been studying light, that may be called akin to it, are conclusive in favor of the mystics, who are the disciples of Saint John the Apostle, and in favor of the great thinkers who have established the spiritual world, in which the relations between man and God are revealed.[19] (my emphasis)

Balzac considered animal magnetism a revolutionary new science that would prove that thought was a kind of imponderable fluid akin to electricity and magnetism. He feels compelled to show his long-standing familiarity with the "wonders" of animal magnetism by dating it back to 1820, the year of Oersted's discovery of electromagnetic interaction. Balzac looked for and studied scientific models that could help him justify his literary inductive method. He was intrigued by Gall's phrenology and Lavater's physiognomy and the way they attempted to unveil invisible traits of the mind by collecting data about the shape of one's face and skull.[20] However, as this crucial passage from the avant-propos shows, Balzac was most impressed with the "miracle" of animal magnetism. He closely studied publications on the subject and mentions in his personal letters that it had cured his sister.[21] According to Théophile Gautier, he even practiced magnetism himself and used somnambulists to seek perceptions ranging from the physical and emotional states of his loved ones to the location of buried treasure.[22]

Magnetic somnambulism brought to the fore telepathic and haunting experiences that appeared to operate beyond the traditional constraints of space and time. Magnetic societies with religious affiliations conducted experiments that went beyond diagnosing illnesses. Nicole Edelman writes, "In their hands, somnambulism and somnambulists become a strange theological tool, a means to investigate extraterrestrial worlds, a machine to solve metaphysical questions."[23] As seen in the case of Friederike Hauffe, these practices also initiated a new literary genre. Many magnetic societies turned somnambulists' clairvoyance into a probe to discover hidden knowledge, from the divine to the practical, carefully recording their words in writing. A foremother of surrealist automatic writers, the somnambulist Hauffe would sometimes write her own texts.[24] Edelman mentions Marie-Louise de Vallières de Monspey, who, starting in 1785, wrote a series of pamphlets on wide-ranging subjects with such titles as "Exposition of the Universal Œuvre," "The Doctrine of Truth," "Of Matter and Forms," "The Knowledge of Beings in Nature," "Plants and Their Properties," and "Theory of the Air Principle or Magnetism."[25] The output of this pioneer somnambulist-writer shows that her interests included both theological and scientific inquiries.[26]

La comédie humaine contains scenes of somnambulist-writing, including at the end of *Louis Lambert*, where the eponymous character is in a constant somnambulic state and dictates philosophical aphorisms to his devoted lover, Mademoiselle Villenoise. Before becoming a somnambulist-writer, Lambert's ceaseless reading paved the way for his state of heightened perception: "As he concentrated all his forces in the book he was reading, he lost to some extent the conscience of his physical life, and did not exist any more but by the very powerful play of his interior organs of which the range had boundlessly extended: as he said, *he left space behind him*" (Balzac's emphasis). Reading also transports Lambert through time: "While reading the account of the battle of Austerlitz [. . .] I saw all of its events. The gun shots, the cries of combatants resounded within my ears and shook my entrails; I was smelling the gunpowder, I was hearing the horses and the men's voices; I was admiring the plain where armed nations were fighting, as if I had been on the height of Santon." Like the magnetic medium of the somnambulist, text allows Lambert to perceive beyond the limits of space and time. At fifteen, following years of constant reading, Lambert's augmented sense of perception reaches new heights when he experiences his first premonition about an event that happens the next day. He concludes, "If without stirring I traversed wide tracts of space, there must be inner faculties independent of the external laws of physics."[27] For Balzac, animal magnetism rendered manifest a deeper domain of interconnection that eluded the physics of his time.

The most significant scene of somnambulist-writing in his work appears in *Ursule Mirouët* (1842). Balzac dedicated this novel in great part to animal magnetism and its history. Doctor Mirouët, the uncle and adoptive father of Ursule Mirouët, owes his fame to his involvement in "the lively debates to which mesmerism gave rise" before the French Revolution, which pitted Mesmer's followers against the medical establishment. Due to his staunch antimesmerist stance, he sacrifices his friendship with Doctor Bouvard, who had become promesmerist. Many years after their bitter separation, and at Bouvard's request, Mirouët decides to give him one last chance to prove, in Bouvard's words, "that magnetism is about to become one of the most important of the sciences."[28]

The two doctors enter a small Parisian apartment where the magnetizer has been waiting for them with his somnambulist, a woman, who seems "to belong to an inferior class" and appears to be sleeping. The magnetizer puts the doctor and the somnambulist in a kind of "rapport" by having him hold her hand. Mirouët tests her clairvoyance by asking a series of personal questions about his home in Nemours. She answers them correctly. In a second séance, Mirouët sets up the ultimate test for the existence of "sympathies" by writing down the somnambulist's testimony about a prayer being recited by Ursule at the same time in Nemours. He ends the séance and rushes back to Nemours with the written proof. It takes him hours to cross the space the somnambulist crossed instantaneously before he finds out that she was right again.

Like Puységur's servant and Hauffe, the somnambulist in *Ursule Mirouët* develops more elegant speech in her state of clairvoyance, prompting Mirouët to say, "A woman of the people to talk like this!" Bouvard replies, "In the state she is in all persons speak with extraordinary perception."[29] Balzac notoriously struggled with his style and how best to convey his thoughts. The elevated speech and clairvoyance of the somnambulist must have caught his attention as a way to develop his own writing style. This is particularly manifest in the transcript of the somnambulist's clairvoyant description of Nemours:

> Entering by the steps which go down to the river, there is to the right, a long brick gallery, in which I see books; it ends in a singular building,—there are wooden bells,

> and a pattern of red eggs. To the left, the wall is covered with climbing plants, wild grapes, Virginia jessamine. In the middle is a sun-dial. There are many plants in pots. Your child is looking at the flowers. She shows them to her nurse—she is making holes in the earth with her trowel, and planting seeds. The nurse is raking the path. The young girl is pure as an angel, but the beginning of love is there, faint as the dawn—.[30]

The structure of this passage in many ways resembles the opening pages of *Le père Goriot*, where the all-seeing narrator skims from the outside to the inside of the boarding house and ultimately peeks into the minds of its residents. Balzac uses the somnambulist as a mouthpiece for a rather typical realist description, with the scene's detailed rendering of objects worthy of many passages in *La comédie humaine*. Due to animal magnetism, the somnambulist speaks like a realist writer when she describes distant places and nonintrusively "reads" Ursule's heart as would Vautrin or, for that matter, any reader of the novel. Balzac's novels represent but also aspire to provoke a state of perception akin to the somnambulist's clairvoyance and Louis Lambert's heightened reading experience. Would it not then be more appropriate to call Balzac's oeuvre *magnetic realism*?

Göran Blix has corroborated the legitimacy of this question. For him, *La comédie humaine* should be considered as a work of "magnetic realism" because Balzac's conception of literary visualization is deeply rooted in Mesmer's fluid mechanics.[31] Mesmerists claimed that they could telepathically influence patients via a "magnetic fluid" channeled by their willpower. Balzac appropriated this theory because it suggested that moral and physical forces could interact and were therefore related. It also matched his economical notion of willpower, since in addition to being emitted, Mesmer's magnetic fluid could also be accumulated and concentrated.

Blix focuses on Balzac's related references to "magnetic gaze" and "second sight" to show how clairvoyant visions are more than mere optics. They represent the ebb and flow of the primordial magnetic fluid as it marks and passes through the physical surface of things. The visible in Balzac is permeated by a kind of magnetic "force field" that imbues it with the products of invisible mental processes such as memory and the contingent valorization associated with desire. Since writing can describe or convey the fluctuations and concentrations of such invisible mental processes, the visualization it evokes is akin to the somnambulist's second sight.

From the way Blix writes about "the occult roots of realism," there is an undercurrent to some of his arguments that actually tones down the impact of the occult in *La comédie humaine*. According to Blix, the aforementioned somnambulist's description of Nemours in *Ursule Mirouët* shows that "the gaze that Balzac brings to bear on the domestic interior *is* the magnetic gaze of the medium." But he minimizes the somnambulist's acumen in the next sentence: "We see here the decisive innovation that [Balzac] brings to occult seeing, namely that he anchors it firmly in the here-and-now, and applies it directly to exposing the secret recesses of contemporary life."[32] Blix convincingly demonstrates how the magnetic somnambulist's "occult seeing" completely informs the supporting structure of *La comédie humaine* and consequently should be considered as one of the main driving forces behind the literary and conceptual innovations of Balzac's realism. However, how Balzac brings his own "decisive innovation" to "occult seeing" or to the roots of realism remains to be proven.

Edelman has traced the history of magnetic somnambulists, who since the mid-1780s—long before Balzac was born—had already described in writing what they had perceived through "second sight." These somnambulists-writers were reportedly endowed with the gift of elevated speech and were already in the habit of probing "the secret recesses of contemporary life" in works on subjects addressing everything from mystical to practical knowledge. What they described was charged with memory, affective energy, contingent valorizations—or what Balzac referred to as "magnetic fluid." When Balzac wrote in *Ursule Mirouët* a detailed description generated through "second sight," he did not impose his style or decisive innovation on it but most likely provided a fairly accurate rendering of the way he thought a "woman of the people" could have spoken in a state of magnetic somnambulism. If anything, through the somnambulist's description of Nemours, Balzac pays his debt to the elevated speech of the women he believed to possess the occult power of second sight and who had supplied the matrix for his own writing.

Balzac and Animal Electromagnetism

Blix compares *La comédie humaine* to a "magnetic radiography" because Balzac can perform a nonintrusive scan of French society due to a monist theoretical framework in which, like two ripples in the same all-pervasive magnetic pool, his "magnetic gaze" and the "magnetic charges" it records participate in the same substance.[33] However, the idea of magnetic monism proved impossible to sustain, especially after Oersted, Ampère, and Faraday showed during the 1820s and 1830s that magnetism was not an independent force, since it could be induced by an electric current.

In *Louis Lambert*, the sudden realization that he possesses second sight prompts Lambert to compile his thoughts in a book of revelations called "Treatise on the Will." For the narrator, this breakthrough event originated "just as the electric sensation always felt by Mesmer at the approach of a particular servant was the starting-point of his discoveries in magnetism." As mentioned above, the experience that unveils Mesmer's animal magnetism and Lambert's second sight is akin to electromagnetic induction: a moving magnet (the approaching servant) induces an electric current in a conductor (Mesmer). Published soon after Faraday's discovery of induction, the revised version of *Louis Lambert* presents an electromagnetic account of animal magnetism that underscores that the "magnetic radiography" of *La comédie humaine* cannot function without electricity.

This unprecedented analogy with electromagnetic induction seems almost too advanced for the 1830s. Yet subsequent occurrences in Balzac's novels confirm that he was indeed familiar with Faraday's and Ampère's works. A year after the revised edition of *Louis Lambert*, it resurfaces in *Séraphîta* (1834), another philosophical novel where Balzac develops his physical and metaphysical ideas. The magnetic and electric mysteries of Mary Shelley's *Frankenstein* come to mind when, in Norway and under the influence of a strange "Polar" climate, the character Minna feels "electric shocks like those of the torpedo" after gazing into the "magnetic eyes" of the androgyne Séraphitüs-Séraphîta.[34] Later, Séraphitüs-Séraphîta compares the power emanating from the soul (and responsible for "sympathies") not to mere magnetism but to an electromagnetic effect: "You believe in the power of the electricity which you find [*fixée*] in the magnet and you deny that which emanates from the soul."[35] "Sympathies" are telepathically transmitted through space, just as the electricity contained in a magnet invisibly induces an electric current in a nearby conductor.

By implying that electricity can magnetically travel through space from one conductor to another, the quotes from *Louis Lambert* and *Séraphîta* also suggest that electricity precedes magnetism. Balzac would then allude more specifically to the theory of André-Marie Ampère, who in 1820, immediately after witnessing Oersted's experiment in Paris, laid the foundation of electrodynamics by arguing that magnetism could be reduced to loops of electric current.[36] Balzac invokes and vulgarizes this theory when he refers to the electricity contained in a magnet and extends this relation to the invisible emanations of his extraordinary characters.[37]

In *Théorie de la démarche* (1833), Balzac argues that an "invisible fluid" akin to Mesmer's and emanating from one's "gait" (*démarche*) should be the object of a science that would provide a privileged access into thought, since they both participate in movement. He describes his initial obsession with this new field of study in relation to Ampère's own distraction.[38] As Ampère's fame grew, he became a household name symbolizing the apparent absent-mindedness associated with natural philosophers consumed by the secrets of nature.

Balzac also drew from Ampère's reputation for Balthazar Claës, the tragically distracted main protagonist of *La recherche de l'absolu* (first published in 1834).[39] At one point, Claës theorizes about mental capacity in terms of an electrical apparatus powered by "electro-magnetism," positing that "the brain of an idiot contains too little phosphorous or other product of electro-magnetism, that of a madman too much; the brain of an ordinary man has but little, while that of a man of genius is saturated to its due degree. The man constantly in love, the street-porter, the dancer, the large eater, are the ones who disperse the force resulting from their electrical apparatus."[40] Balzac expands electromagnetic interaction to organic life by suggesting that phosphorous, which many believed to be closely linked to vital forces at the time, was one of its products. Genius, willpower, desire, and appetite fundamentally depend on this electromagnetic apparatus.[41]

The characterization of *La comédie humaine* as a kind of "magnetic radiography" implies that Balzac championed a monist interpretation of occult interactions and that his magnetic realism simply records the magnetic emissions of various social motors without undergoing any transformation or conversion processes. Emerging sporadically in his philosophical works of the 1830s, Balzac's prose registers a conceptual change triggered by a new transformational motor powered by electromagnetic interaction. In its most sophisticated moments, the "magnetic gaze"—which turns extraordinary figures like Vautrin, Louis Lambert, Séraphîtüs-Séraphîta, and Mesmer into clairvoyant mind readers—works through inductions that continuously convert the electric into the magnetic, and vice versa. Through realist descriptions reminiscent of those of the somnambulist-writer, *La comédie humaine* emulates such an induction apparatus by aspiring to provoke in the reader the experience of "second sight." Along with the "miracles" of animal magnetism, electromagnetic induction offered Balzac a physical model to represent and complicate the nature of elusive interconnections such as "sympathies."

Since his prose relies less on masking irony, animal electromagnetism emerges much more transparently in Balzac than it does in Poe. Although it remains in a nascent stage, its metonymic function is already apparent. Alongside the occult operations of magnetic monism, there emerges an even stranger model of interconnection based on the relation of contiguity revealed by electromagnetic transformations. The unprecedented appearance of such electromagnetic imagery in *Louis Lambert* also shows that Balzac was aware of its epistemological significance for thought. His account of the process of discovery establishes a link between Faraday's induction apparatus and inductive reasoning that signals a reevaluation of what comes into play when scientists and encyclopedic writers like himself unveil new principles. Balzac's rapprochement between the two kinds of induction marks a stepping-stone in the emergence of electromagnetic thinking that becomes particularly significant in light of contemporary interpretations of these two influential applications of the term *induction*.

The Term *Induction* in Electrical Science

Faraday begins the series of papers describing the experiments that guided him to the discovery of electromagnetic induction by justifying his terminological choice: "The power which electricity of tension possesses of causing an opposite electrical state in its vicinity has been expressed by the general term Induction; which, as it has been received into scientific language, may also, with propriety, be used in the same general sense to express the power which electrical currents may possess of inducing any particular state upon matter in their immediate neighbourhood, otherwise indifferent. It is with this meaning that I purpose using it in the present paper."[42] The word *induction* initially referred to electrostatic phenomena. Unlike conduction, or charging by contact, *induction* refers to the production of an opposite charge through space. A positively charged body simultaneously gives rise to a negatively charged one, and vice versa. Although they relied on different terminologies, Benjamin Franklin (1706–1790), John Canton (1718–1772), Johan Carl Wilcke (1732–1796), and Franz Aepinus (1724–1802) are credited by historians with the first formulation of the concept of electrostatic induction.[43] In 1800 Alessandro Volta (1745–1827) invented the battery, which revealed the different electrical phenomena dependent on the steady generation of a current. Faraday proposed retaining the general meaning of *electrostatic induction* to describe similar influences observed in electric currents but also expanded its use to describe any state arising through these currents' affinity with magnetism.

From the second half of the eighteenth century to the beginning of the nineteenth, *induction* progressively made its way into the official terminology of electrical science. In 1777 Tiberius Cavallo (1749–1809) stated in his treatise on electricity, "The action of these plates depends upon the principle long ago discovered, *viz.* the power that an excited electric has to induce a contrary Electricity in a body brought within its sphere of action."[44] Cavallo does not explain his choice of the verb *to induce* for this electrical effect. He follows the verb's typical eighteenth-century dictionary definition of producing or bringing into view by influence or exterior cause. (The same meaning of the verb appears elsewhere in the treatise in the more familiar nonelectrical context.)[45] At times, Cavallo also relies on *to induce* in the Baconian sense.

The introduction of the term *induction* in electrical science did not happen without controversy. In his often-cited 1814 treatise on electricity, George John Singer (1786–1817) writes on the subject of electrostatic induction, "Such phenomena are classed under the general term electrical influence; and positive and negative states so produced are called the electricities of position, or approximation, and by some writers induced electricity."[46] The main writer whom Singer has in mind when he reluctantly mentions the term *induced electricity* is Humphry Davy (1778–1829), the great pioneer in electrochemistry and Faraday's old boss at the Royal Institution.

In 1812 Davy had advocated the use of *induced electricity* and *induction* in his descriptions of electrical effects.[47] In an article predating his treatise, Singer had criticized Davy's indiscriminate use of *induction* to refer to electrical effects that he thought were actually different and stated that "in its literal interpretation [induction] expresses nothing analogous to any known electrical effect."[48] Singer also had his critics. For instance, a reviewer of Davy's work, aware of Singer's terminological objection, stated, "As to the term induction, which is more familiar to metaphysical than physical language, it seems as convenient and applicable as any other."[49] Although the debate over the value of the term *induction* in electrical science would go on throughout the nineteenth century,[50] Davy's usage quickly became the norm.

Whether its literal interpretation fails to convey the nature of electrical effects or is as good as any other, the early controversy surrounding Davy's choice of *induction* manifests a more profound epistemological issue linked at the time to the Newtonian framework of electrostatics. During the eighteenth century, Newton's physics became the go-to tool to unveil the secrets of the universe. In 1785 Charles-Augustin de Coulomb (1736–1806) published the law uncovering the mathematical relation between electrostatic force and the interaction of electrically charged particles. Coulomb's law ($F = kq1q2/r^2$) looks structurally the same as the law of universal gravitation ($F = Gm1m2/r^2$), suggesting that the fundamental principles of Newtonian physics are at work in all natural forces. However, as in Newton's law, Coulomb's law implied a type of action at a distance that occurs without delay or mediation. The actual way in which electricity generated an action through space remained a mystery, and the debate around whether Davy's use of *induction* provided the most accurate term for a kind of electrostatic influence sidestepped the critical issue, since regardless of terminology employed, induction could only refer to a vague action at a distance.

During the first half of the nineteenth century, Coulomb's achievement prompted other natural philosophers to apply Newtonian physics to magnetic and electromagnetic phenomena, although without conclusive success. The most influential attempt came from Ampère. He considered magnetism the product of an electric current and studied electromagnetic effects in terms of the attraction and repulsion of current-carrying wires and action at a distance. The reduction of magnetism to an electric current helped him quantify the relation between the two forces in what is now known as Ampère's law. Confident of the electric nature of magnetism, he rejected Oersted's *electromagnetic* and introduced *electrodynamic action* (as opposed to *electro*static *action*).[51]

The success of Ampère's electrodynamics contributed to the elision of magnetism in the representation of phenomena brought forth by the discovery of electromagnetism. However, while Ampère was laying the foundation of electrodynamics, others such as Johann Joseph Prechtl (1778–1854) and Jöns Jacob Berzelius (1779–1848) were taking the opposite approach as they attempted to explain the electric current in relation to magnetism.[52] Although their failed and forgotten theories lacked the mathematical clarity of Ampère's, they serve as historical reminders that, despite its achievements, electrodynamics remains a convention. As discussed below, Einstein's special theory of relativity relegitimized the use of *electromagnetism* by arguing that, in the phenomenon of electromagnetic induction, electric currents and magnetism are manifestations of the same fundamental entity—the electromagnetic field—and that they appear different only due to the observer's frame of reference.

Faraday applauded Ampère's theory of electrodynamics as one of the best theories available but had doubts concerning its empirical foundation. Following his discovery of electromagnetic induction, Faraday struggled to find a reliable principle that could account for the puzzling interactions of magnetism, motion, and voltaic electricity. He avoided Ampère's terminology because it implied the primacy of the electric current, which for him remained to be proven. Instead of electrodynamic induction, he adopted a nomenclature that did not commit to a specific theoretical preconception. In papers exploring inductive effects through various experimental apparatuses, he distinguishes whether it is an ordinary magnet or a current-carrying wire that is the source of magnetism responsible for inducing electricity by calling the phenomena "volta-electric" and "magneto-electric" induction, respectively.[53]

Faraday's noncommittal approach to Ampère's electrodynamics helped him develop a radically different framework that did not depend on Newtonian action at a distance and that would, as James Clerk Maxwell later showed, prove more appropriate to detect and formalize the general principles of electromagnetism. He shifted his attention to the "magnetic curves" drawn by iron filings around a magnet as a way to visualize and comprehend the nature of electromagnetic influences.[54] He also realized that he could consistently predict the electromagnetic effects of induction by focusing on the way a conductor in relative motion to a source of magnetism "cuts" its magnetic curves.[55] This first formulation of what textbooks now call Faraday's law of induction originated in his skepticism concerning the prevailing electrodynamic theory of his time and his interest in the previously ignored magnetic curves.

Electromagnetic Induction and Inductive Reasoning

Faraday's works on electromagnetic induction not only led to the theoretical leap that brought forth field theory but also helped legitimize alternative methods of discovery based on more open-ended and less detached methods of investigation. In his landmark *Treatise on Electricity and Magnetism* (1873), Maxwell develops a mathematical formalism compatible with Faraday's concept of lines of force and derives from it a version of Ampère's law. Maxwell also urges students to read both Faraday's and Ampère's works because they display two different types of scientific thinking. He calls Ampère "the Newton of electricity" and praises him for his mathematical clarity, but he notes that despite providing experiments confirming his law of action, Ampère nevertheless keeps hidden the process that initially led to its discovery: "The method of Ampère, however, though cast into an inductive form, does not allow us to trace the formation of the ideas which guided it. We can scarcely believe that Ampère really discovered the law of action by means of the experiments which he describes. We are led to suspect, what, indeed, he tells us himself, that he discovered the law by some process which he has not shewn us, and that when he had afterwards built up a perfect demonstration he removed all traces of the scaffolding by which he had raised it." For its part, Faraday's *Experimental Researches* displays an inductive method, which for Maxwell is "better fitted for a nascent science" because it provides a more detailed account of the process guiding his discoveries. Instead of "a perfect demonstration," Faraday candidly shares the trials and errors that led him to his conclusion: "Faraday, on the other hand, shews us his unsuccessful as well as his successful experiments, and his crude ideas as well as his developed ones, and the reader, however inferior to him in inductive power, feels sympathy even more than admiration, and is tempted to believe that, if he had the opportunity, he too would be a discoverer."[56] To Ampère's detached mathematical demonstration Maxwell compares Faraday's personal narrative of his struggles "to coordinate his ideas with his facts."[57] In addition to being innovative, Faraday's detailed account of the process of discovery conveyed what really happens in a lab and thus was more appropriate for "the cultivation of a scientific spirit."[58]

Maxwell's characterization of the contrasting scientific styles of Ampère and Faraday echoes the influential debate on the nature of inductive reasoning that had recently occurred between philosophers of science William Whewell and John Stuart Mill. The inventor of the word *scientist*, Whewell was a friend of Faraday, who often solicited him for advice on naming the new phenomena he was studying. *Anode*, *cathode*, and *ion* were some of Whewell's contributions to Faraday's nomenclature.

Whewell's theory of induction was a mix of empiricism and apriorism, with observation working in tandem with an innate conceptual framework of ideas to unveil the laws and causes of phenomena. Induction establishes a chain of connections where there once was none: "If we consider the facts of external nature to lie before us like a heap of pearls of various forms and sizes, mere Observation takes up an indiscriminate handful of them; Induction seizes some thread on which a portion of the heap are strung, and binds such threads together."[59] Unlike deduction, Whewell argues, induction can reveal new principles through the detection of a fundamental "unity" among what appears completely unrelated:

> The deductive people go on following, illustrating, expanding, a given notion which of the nature of it must be defined and limited and so restricted to the range of our primitive knowledge—But the minds that feel a conviction of principles of unity as yet undetected, that believe in the existence of truths wider than they can limit by phrases habitually current, and that assent to the possibility of a connection among laws that seem far asunder, while they acknowledge their ignorance what the connection is; these are the minds which have the best chance of discovering new principles and new generalizations and such habits of thought lead naturally to the persuasion of a supreme principle of unity and connexion.[60]

For Whewell, the process of induction cannot be reduced to the collection of instances to prove a general principle. What he called "discoverer's induction" establishes connections that involve more than just the metaphoric processes of identifying a resemblance among observations or the deduction of conclusions deriving from the same principle. According to Whewell, "discoverer's induction" should lead from the same to the different or from the known to the unknown. Its most remarkable trait is the detection of metonymic links.

Laura J. Snyder has shown that the debate between Whewell and Mill about inductive reasoning hinged on the former's apriorism. Mill thought that a conceptual framework based on ideas not acquired from experience participated in the school of intuitionism and could only be a source of arbitrariness and prejudice. For Mill, the object of scientific knowledge must derive strictly from experience and logical inference. He therefore argued that Whewell's "discoverer's induction" was not an inductive method because it included nonrational guesswork guided by vague a priori conceptions. Mill proposed instead his own theory of induction, which, he contended, implied no a prioris. Mill's account of induction prevailed, and throughout the twentieth century, philosophers considered Whewell's theory of induction as noninductivist or closer to Peirce's abduction.[61]

Whewell responded to Mill's critique by characterizing him as a deductive thinker. Mill's sole focus on what could be empirically demonstrated limited his method to deriving conclusions from known principles instead of finding new ones. Whewell reaffirmed that the process of induction includes "a mysterious step,"[62] which Mill would deem impossible to demonstrate but which nevertheless occurs and involves the mental act of making an appropriate analogical link, or *colligation*, between empirical observation and an a priori idea.

Whewell and Mill defended their respective stands with examples drawn from the history of science. One of them concerned Faraday and the discovery of electromagnetic induction. Mill used it to illustrate what he calls the "methods of elimination,"[63] which he considers the foundation of inductive reasoning. These methods investigate what produces a phenomenon by varying the circumstances of its occurrence and nonoccurrence and thus identifying, through a process of comparison and elimination, its true cause. For Mill, Faraday's series of experiments on induced electricity continues the exploration of the phenomenon of electrostatic induction. Voltaic current and electromagnetic effects provided new instances to test the nature of the simultaneous apparition of opposite electrical states and identify its cause. Through methods of elimination, Mill concluded that Faraday had discovered a similar cause behind static and voltaic electricity and that what we now describe as electromagnetic induction was instead a phenomenon of a "different class."[64]

Whewell faulted Mill for assuming that Faraday believed in the identity of static and voltaic electricity. He also criticized him for claiming that Faraday's main contribution boiled down to confirming and extending the reach of the principle of opposite electricities, which had been known since the time of Benjamin Franklin.[65] For Whewell, Faraday's most important discovery was in the domain of electromagnetic effects and revealed a previously unknown link between magnetic and electric polarities. Whewell considered the notion of polarity as one of the main a priori ideas framing scientific research. As discussed further in chapter 3, the term originated first in the study of magnetism when, in the thirteenth century, Pierre de Maricourt reinterpreted the attraction and repulsion of magnets in terms of the interaction of their "poles."[66] Over time, the identification of other types of polarity in electrostatics, optics, and chemistry not only brought support to the notion of polarity as an a priori idea; it improved its understanding. For Whewell, Ampère had contributed to this overall improvement with his electrodynamic theory, which reduced the distinct polarities of voltaic electricity and magnetism in terms of the former.[67]

Aaron D. Cobb has recently reassessed the point of contention between Whewell and Mill concerning Faraday's inductive reasoning and the discovery of electromagnetic induction. He agrees with Whewell that Mill's interpretation of Faraday's work and method is wrong and that it mainly serves the purpose of justifying Mill's methods of elimination without displaying much regard for historical accuracy or what actually happened in the lab. Cobb also notes that although Ampère's electrodynamic theory played a guiding role in Faraday's experiments, it was not as crucial as Whewell suggested.[68]

The different views highlighted by the Whewell-Mill debate over the inductive method reverberate in Maxwell's praise of Faraday's inductive method for displaying the struggle "to coordinate his ideas with his facts" and of Ampère's theory for its mathematical clarity. Whewell and Mill devised contrasting logics of discovery that had at least one aspiration in common: the metonymic imperative of leading from the known to the unknown by crossing the threshold between the same and the different. Maxwell's characterization of Ampère recalls the extreme empiricism and rationalism championed by Mill. Such a method leads to purely technical demonstrations that "removed all traces of the scaffolding by which [the author] had raised it."[69] For their part, Faraday's detailed narratives showing the constant interaction of ideas and facts evoke Whewell's argument that inductive reasoning embodies a bridge between empiricism and apriorism that involves a nondemonstrable "mysterious step," which he associated with the analogical process of colligation. Maxwell does not hide his preference for Faraday's inductive method and particularly appreciates the way it presented a more authentic account of the process of scientific discovery.

Scholars continue to differ widely on the main methodological factors that led Faraday to his discoveries. Like Mill, some have portrayed him as mainly an empiricist, while others have followed Whewell, contending that a priori ideas and metaphysical elements also played a crucial role.[70] Friedrich Steinle has recently described his approach in terms of "exploratory experimentation":[71] Instead of designing experiments to test a preestablished idea or theory, Faraday systematically varied experimental parameters to reduce inductive effects to their essential features. Once this empirical reduction was achieved, Faraday realized that these features did not comply with existing concepts and categories, and he proceeded to revise them by developing a radically new theoretical framework based on the idea of "magnetic curves." Faraday's exploratory experimentation highlights the effectiveness of a methodological approach that depends much more on process than on theory. Although the variation of experimental parameters is systematic, its main purpose is not to confirm theoretical expectations. The outcome of this process remains both more open-ended and more attuned to the need for conceptual change.

Steinle's description of exploratory experimentation downplays other factors that constitute an integral part of the process of discovery. Balzac recognized in electromagnetic induction a relation of contiguity that prompted him to revise his notion of how inductive reasoning works. In *Louis Lambert*, his initial analogy with Newtonian physics conveys a straightforward experience where the detached scientist discovers universal gravity by witnessing the fall of a fruit. By replacing gravity with electromagnetism, Balzac invokes an induction apparatus that represents a more complex experience, where Mesmer's discovery of animal magnetism proceeds indirectly via an "electric sensation." Here the body of the observer intimately participates in detection. Unlike Mill's extreme empiricism and Steinle's exploratory experimentation, Balzac's account underscores the role played by the body and offers a more all-encompassing take on the method of discovery.

Balzac refers to electromagnetic induction to highlight an indirect process in inductive reasoning that resonates more with Faraday's narratives of his trials and errors and Whewell's "mysterious step" than with Ampère's direct (or scaffolding-free) demonstration. As Dupin will later argue in his account of metonymic induction, the discoverer must be "head" and "body." As an open-ended process, exploratory experimentation involves decision-making and theoretical leaps deriving from both conscious and nonconscious influences. Balzac was the first to perceive in electromagnetic induction a new and more accurate model to represent the volatile aspect of inductive reasoning and to consider nonconscious phenomena rendered manifest by magnetic somnambulism central to understanding this method of discovery.

Einstein's Induction Apparatus

Balzac mobilized the relation of contiguity unveiled by Faraday's induction apparatus to rethink inductive reasoning and literary realism long before the emergence of field theory. To demonstrate the significance of his early engagement with electromagnetic contiguity, I will show how his engagement anticipates Einstein's. Although both luminaries relied on the concept of electromagnetic contiguity to undermine the Newtonian conception of space and time, their works are obviously different. I will therefore briefly discuss the scientific breakthrough Einstein achieved with his special theory of relativity before examining the crucial role electromagnetic contiguity played in its discovery.

The development of field theory during the second half of the nineteenth century brought physics into a state of crisis. Whereas Maxwell's electromagnetic laws applied to the continuous and imponderable domain of the ether, Newton's mechanical laws of motion ruled over the discontinuous and atomistic domain of matter. Field and matter represented opposing views about the nature of the universe that threatened to undermine the unity of physics. Attempts to reconcile them through various ether theories struggled to explain how the physics of continuity and discontinuity interacted.[72]

Before Einstein came to the fore, physicists thought that, similarly to ocean waves, light and other electromagnetic radiations were undulations of various sizes spreading through a medium permeating space commonly known as the *ether*. Fundamental questions remained concerning the exact nature and structure of the invisible and all-pervasive electromagnetic ether. Was it material, immaterial, or a different kind of matter? How did the elusive ether interact with material objects? How did the imponderable connect with the ponderable?

Experiments designed to study the relation between the ether and matter yielded disturbing results. The most famous, the Michelson-Morley experiment, took place in 1887. It built upon the premise that because Earth was moving with great velocity through the all-pervasive electromagnetic ether, an ether wind should be detected on the surface of the planet. As an electromagnetic wave, a beam of light would then travel at different speeds depending on whether it was pointed along or against the ether wind. The results of the experiment did not show any difference. The speed of light remained constant in all directions. No ether wind appeared to brush across the earth. The ether resisted detection. The unchanging speed of the light wave also implied that the ether might not even exist or that it might be reminiscent of a Newtonian absolute space. Physicists held on to the idea of the ether and developed more intricate theories to account for the absence of ether wind, but experimental validation continued to elude them.[73]

In 1905 Einstein published "On the Electrodynamics of Moving Bodies," the paper that would render the idea of the ether superfluous by refounding physics on the new grounds of his special theory of relativity. The special theory of relativity consists of two postulates. The first, derived from Galilean relativity, stipulates that the laws of physics are the same for observers in all inertial frames of reference. For instance, let us consider two different inertial frames of reference: (1) the platform of a train station and (2) a train moving in a straight line at a constant speed. The observer on the station platform measuring the speed of a ball thrown from one boy to another will come up with the same result as the observer inside the train measuring the exact same action taking place within his car. If the observer on the platform measures the same ball toss taking place in the train passing by him, the speed of the ball will be different, since from his perspective, he has to add the speed of the train to that of the ball to get the correct result. The speed of the ball is therefore not absolute; rather, its value is relative to the reference frame of the observer. Galilean relativity had also been instrumental in the elaboration of Newton's laws of mechanics. Yet the absolute space where action at a distance took place was an oversimplification of the principle of relativity because it implied the existence of a fixed frame of reference. By raising the principle of relativity to a fundamental law of physics, Einstein undermined the Newtonian idea of absolute space by making such a privileged frame of reference theoretically impossible and by shifting the focus of physics to the way reference frames *relate* to each other.

The second postulate of the special theory of relativity modifies Galilean relativity with the particular case of light: the speed of light in a vacuum is the same for all observers regardless of their frames of reference. For the train station observer, a beam of light fired inside the train passing by him has the same speed as a beam of light fired on the platform. Unlike the boy's ball, the speed of light is not affected by the speed of the train. Although the failure to detect any variation in the speed of light in experiments such as Michelson-Morley's seems to provide some ground for this postulate, it remains a bold claim that would still puzzle most observers today.

The two postulates of the special theory of relativity yielded counterintuitive insights about the nature of the universe. According to Einstein's second postulate, for the observer on the station platform, the speed of a beam of light is not affected by the speed of the train in which it has been fired. The speed of the beam of light must then influence the spatiotemporal conditions of its own measurement: for the speed of light to remain constant for all observers regardless of their reference frames, time and space cannot be absolute.

By applying his second postulate to simple thought experiments, Einstein predicted that a clock in motion, especially as it approaches the speed of light, would tick at a slower pace and shrink relative to a stationary clock. Known as *time dilation* and *length contraction*, these strange consequences of the special theory of relativity and their eventual experimental confirmation during the twentieth century marked the end of the notions of absolute time and space, which had structured the theoretical framework of Newtonian classical mechanics. Whereas the concept of Newtonian mechanics works well with objects moving at a low speed, Einstein showed that at a higher speed it becomes unreliable. With simple mathematical adjustments prompted by the newly discovered nature of time and space, Einstein's relativist mechanics proved to be a more trustworthy theory that could account for all speeds (the speed of light being the ultimate speed). As the more accurate and all-encompassing theory, Einstein's special relativity displaced Newtonian physics and remains to this day one of the fundamental principles of physics.

Einstein considered electromagnetic induction the key to understanding the inductive reasoning that led him to the discovery of special relativity and later "general relativity." Before introducing his two postulates, Einstein begins "On the Electrodynamics of Moving Bodies" with a general description of the relation of contiguity unveiled by Faraday's 1831 induction apparatus. As Faraday showed, the induced current depends on the interaction of a source of magnetism and a conductor: "It is well known that Maxwell's electrodynamics—as usually understood at present—when applied to moving bodies, leads to asymmetries that do not seem to attach to the phenomena. Let us recall, for example, the electrodynamic interaction between a magnet and a conductor. The observable phenomenon depends here only on the relative motion of conductor and magnet, while according to the customary conception the two cases, in which, respectively, either the one or the other of the two bodies is the one in motion, are to be strictly differentiated from each other."[74] Electromagnetic induction involves both mechanics (the motion of bodies) and electromagnetism. Einstein notes that the Maxwellian interpretation of electromagnetic induction makes a theoretical distinction between whether it is the conductor or the magnet that moves. However, for Einstein this distinction must be artificial because the induced electric current only depends on the relative motion of the conductor and the magnet.

In later writings, Einstein provides a more detailed account of the reasoning that prompted him to apply the principle of relativity, a fundamental law of mechanics, to electromagnetic induction: "Up to [the special theory of relativity] the electric field and the magnetic field were regarded as existing separately even if a close causal correlation between the two types of field was provided by Maxwell's field equations. But the special theory of relativity showed that this causal correlation corresponds to an essential identity of the two types of field. In fact, the same condition of space, which in one coordinate system appears as a pure magnetic field, appears simultaneously in another coordinate system in relative motion as an electric field, and vice versa."[75] Before special relativity, physicists had assumed that electric and magnetic fields constituted two distinct domains. They considered them as sovereign fields that only shared a "causal correlation." According to special relativity, however, the electric and magnetic fields are two manifestations of the same entity—the electromagnetic field. The difference between the electric and magnetic fields now derives from the reference frame (or "coordinate system") of the observer:

> The difference between [the electric and magnetic fields] could be not a real difference, but rather, in my conviction, could only be a difference in the choice of reference point. Judged from the magnet there certainly were no electric fields; judged from the conducting circuit there certainly was one. The existence of an electric field was therefore a relative one, depending on the state of motion of the coordinate system being used, and a kind of objective reality could be granted only to the electric and magnetic field together, quite apart from the state of relative motion of the observer or the coordinate system. *The phenomenon of the electromagnetic induction forced me to postulate the (special) relativity principle.*[76] (my emphasis)

Einstein's second postulate also appears to have been an outcome of the application of the principle of relativity to electromagnetic induction. In this account, the nature of the electromagnetic field recalls the new status of the electromagnetic wave known as light. The electric and magnetic fields are relative manifestations of an objective reality (the electromagnetic field) whose nature must consequently remain the same for all observers regardless of their reference frame. In his reexamination of electromagnetic contiguity, Einstein reconceptualized its metonymic relation and difference in terms of Galilean relativity.

In 1907, two years after publishing his special theory of relativity, Einstein had "the happiest thought of [his] life" when he made an analogy between the gravitational field and his relativist interpretation of electromagnetic induction: "Just as in the case where an electric field is produced by electromagnetic induction, the gravitational field similarly has only a relative existence. Thus, for an observer in free fall from the roof of a house there exists, during his fall, no gravitational field—at least not in his immediate vicinity. If the observer releases any objects, they will remain, relative to him, in a state of rest."[77] As with the electric and magnetic fields, the gravitational field is relative. Einstein's "happiest thought" marked the beginning of years of work that culminated in 1916 with the inclusion of gravity in the theory (or general theory) of relativity.

Einstein reveals how central electromagnetic contiguity was to his thought process. Since Faraday's induction apparatus, electromagnetic contiguity had provided scientists with a fresh empirical framework that helped them reconceptualize the spatiotemporal fabric of the universe through the development of both field theory and the theory of relativity. Fascinated by "sympathies" that contradicted traditional conceptions of space and time, Balzac pioneered the literary appropriation of Faraday's induction apparatus to represent and explore other relations of contiguity he identified in the experiences of reading and inductive reasoning. Like Emerson, Balzac participated in the Faradayan "wave" that swept the nineteenth century and culminated in the theory of relativity. But he was part of a less apparent yet deeper undercurrent that connected him to Einstein based on their common recognition of electromagnetic induction as a new transformational motor for conceptual change and innovation.

3.

Automata

Electromagnetic Interaction After 1850

Although scholars have mainly focused on the emergence of field theory as a unifying concept between Einstein's works and nineteenth-century literature, I argued in chapter 2 for an older and more fundamental point of contact between Balzac and the great physicist based on their mutual use of Faraday's induction apparatus and its display of electromagnetic contiguity to undermine traditional conceptions of space and time. This link between Balzac and Einstein has been overlooked because electromagnetic induction quickly gave way to world-changing technological and conceptual inventions—developments that built upon its power while rendering its mode of operation less and less apparent. As Halliday, Gilmore, and Lieberman have demonstrated, during the second half of the nineteenth century, telegraph and telephone lines, the dynamo, and the electric grid became the go-to analogies for representing interconnection. These technologies manifested metonymic and metaphoric phenomena that materialized through devices different from Oersted's and Faraday's. Yet from Balzac to Einstein, electromagnetic contiguity remained an important motor for conceptual exploration and invention. Tracing some of the legacies of this conceptual motor, I hope to show how it remained significant during the era that separated the two luminaries and that witnessed the emerging prominence of the telegraph and power grid.

Electromagnetic analogies like those found in Poe and Balzac continued to appear during the 1850s and 1860s, and Halliday has noted their presence in the novels of Herman Melville (1819–1891). Drawing inspiration from his friend Nathaniel Hawthorne's electric imagery and reference to "the magnetic chain of humanity" in the short story "Ethan Brand" (1850), Melville relies on electromagnetic interactions in *Moby Dick* (1851) and *Pierre* (1852) to represent "intersubjective *reciprocity*" and "*hierarchy*" (Halliday's emphasis).[1] In France, Comte de Lautréamont (1846–1870) invokes them in book 5 of *Les chants de Maldoror* (1868–69), where the eponymous character exerts a power of fascination described as "being magnetized by the electricity of the unknown."[2]

More common are analogies that invoke familiar technologies without necessarily calling attention to the electromagnetic interaction powering them. In 1871 Arthur Rimbaud writes several letters where he develops his conception of modern poetry in a way that is reminiscent of Poe's notion of psychal impressions and Balzac's electromagnetic realism. Yet Rimbaud relies on Morse's telegraph as his main electromagnetic apparatus to support his vision of poetic interconnection. He claims that a poet must make "himself a *seer* by a long, gigantic, and rational *derangement* of *all the senses*." This method will lead to a language that "will be of the soul for the soul, containing everything, smells, sounds, colors, thought holding on to thought and pulling."[3] On August 15 of the same year, Rimbaud sent a letter to Théodore de Banville in which he describes this language in the poem "What Is Said to the Poet Concerning Flowers" ("Ce qu'on dit au Poète à propos de fleurs"):

> [. . .] médium !
> [. . .]
>
> De tes noirs Poèmes, – Jongleur !
> Blancs, verts, et rouges dioptriques,
> Que s'évadent d'étranges fleurs
> Et des papillons électriques !
>
> Voilà ! c'est le Siècle d'enfer !
> Et les poteaux télégraphiques
> Vont orner, – lyre aux chants de fer,
> Tes omoplates magnifiques
>
> ([. . .] medium!
> [. . .]
> From your black Poems,—Juggler!
> White, green, and red dioptrics,

Let strange flowers burst forth
And electric butterflies!

There now! it is the Century of hell!
And the telegraph poles
Will embellish,—lyre with iron voice,
Your magnificent shoulder blades!)[4]

Calling for a renewal of poetic arts, these verses come from a poem that mocks and parodies the "flowers" of classic poetry. They are addressed to a kind of seer, a "medium," and urge him to write poetry that generates "electric" imagery.[5] To create such wonders, the modern poet must replace the ancient strings of his lyre with the conducting wires of "telegraph poles."[6] With this new, electromagnetic lyre, the modern poet can aspire to a language "of the soul for the soul." Like *Ion*'s magnetic chain, this telepathic language overcomes the mediation of words. As in Balzac's and Poe's explorations of animal electromagnetism, Rimbaud brings together the figures of the poet, the seer, and an electromagnetic apparatus (the telegraph) to conceive of a kind of literary experience that surpasses mere mimesis. But the relation of contiguity between electricity and magnetism has now become less apparent, taking second stage to the telegraph's conducting wires.

A notable exception to the trend of deemphasizing electromagnetic contiguity is Villiers de l'Isle-Adam's *L'Ève future* (*Tomorrow's Eve*, 1878–86), where electromagnetic interactions as well as technologies are central to the narrative and its meditation on the nature of life and cognition. *L'Ève future* is a *roman d'anticipation* dedicated to Thomas Alva Edison (1847–1931), the great engineer and practical scientist who invented the phonograph and who had recently deployed the dynamos that began to deliver electricity and light to the masses. After he immortalized sound and tamed electromagnetism with his new and/or reworked principles and their ingenious applications, the press nicknamed Edison "the Wizard of Menlo Park." In *L'Ève future*, the great "Wizard" of electricity is also a powerful magnetizer. Villiers's portrayal of the charismatic inventor recalls other great mesmeric characters, such as Vautrin and Dupin. Edison is different from them, however, because he can also rely on his electromagnetic machines to telecommunicate. Yet unlike Rimbaud's letter, the way the novel represents animal electromagnetism through Edison's famous invention continues to emphasize how they were powered by the electromagnetic interaction that had fascinated Balzac and Poe. I thus focus on *L'Ève future* to explore animal electromagnetism's legacies, its technological reincarnations during the second half of the nineteenth century, and its continued role as a stimulus for the metonymic imagination.

When Villiers exploits the technological and scientific context of Edison's inventions to render natural what at some points seems supernatural, he compiles a nineteenth-century encyclopedia of (electro-)magnetic tropes. A close examination of these tropes will show how Villiers relied on them to imagine an automaton capable of replacing a human being. Like many of his contemporaries, he saw in electromagnetic motors a way to achieve more "life-like" effects (fig. 9). Building upon Georges Canguilhem's critical examination of the long philosophical and scientific traditions of explaining living beings in terms of self-moving machines, I argue that with the fictional fabrication of a "new Eve," Villiers ironically imitates this tradition in order to debunk it. He regarded such a mechanical conception of the living as reductive and as a negative outcome of the bourgeoisie's rise to power during the Industrial Revolution. By combining spatial and temporal metonymic interconnections rendered manifest by electromagnetism and hypnotic phenomena linked to animal magnetism and mourning, the fictional Edison creates a kind of electromagnetic animal that brings forth a new conception of life as it undermines its mechanical interpretation.

Villiers's conceptual innovations have been important influences on science fiction and theoretical discussions concerning artificial life and intelligence.[7] In his intellectual and cultural history of the automaton, Minsoo Kang contrasts the novel's "new Eve" with previous Romantic examples, such as Olimpia in E. T. A. Hoffmann's "The Sandman" (1816).[8] Although both characters imitate human beings, Olimpia is not alive. As Freud and others have demonstrated, this automaton sheds light on cognition through its uncanny relationship with the character Nathanael who, under the sway of traumatic reminiscences and unable to distinguish between an animate and inanimate object, has fallen in love with the machine. In *L'Ève future*, Edison's automaton has come to life with the help of a magnetic somnambulist who has "incorporated" herself into the machine. Part human and part machine, the "new Eve" is a hybrid entity that explores the nature of life and cognition by blurring the line between the biological and mechanical. Such amalgamations proliferated during the Industrial Revolution due to the mid-nineteenth-century development of the theory of thermodynamics, which, through its unifying concept of energy, rendered

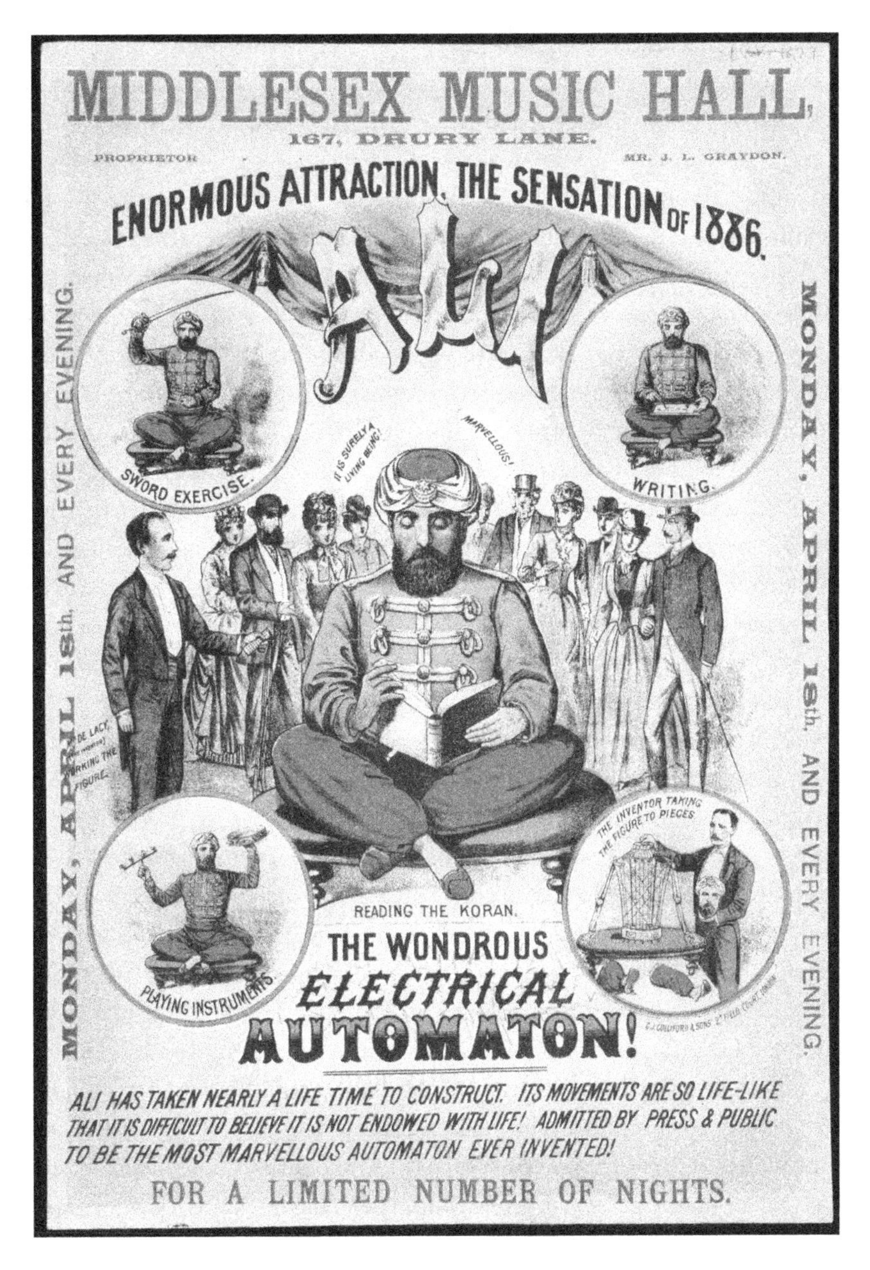

Fig. 9 This poster advertises a "wondrous electrical automaton" named Ali that has fascinated audiences since 1886, the same year Villiers completed *L'Ève future*. Showcased around the world, Ali was part of a new generation of sensational automata that allegedly performed extraordinary feats due to electromagnetic power. Along with its ability to read, write, wield a sword, and play instruments, the poster trumpets, Ali's "movements are so life-like that it is difficult to believe it is not endowed with life!" Photo © The British Library Board / The Image Works.

human and machine work scientifically equivalent.[9] The steam engine played a central role in establishing the principles of thermodynamics and, for Kang, is emblematic of the new generation of hybrid automata spearheaded by *L'Ève future*.

Yet the automaton Villiers invented to reconsider the nature of life and cognition is powered by a different transformational motor. Kang's focus on the steam engine overlooks the more specific contributions of electromagnetic interaction that I have examined thus far and that Villiers appropriated for his own purpose. Literary scholars have also tended to brush over the related scientific and symbolic significance of electromagnetism in *L'Ève future*. Shifting critical attention from the steam engine to the electromagnetic motor clarifies the latter's function and uncovers other important epistemological shifts that did not derive simply from thermodynamics and its conceptions of energy, entropy, and heat death.[10] Such a shift also resituates Villiers's automaton within an older and influential line of metonymic thinking: an alternative model to the mechanical interpretation of the universe that relied on self-moving machines powered by magnetic bipolarity.

Identified during the thirteenth century, the bipolar structure of magnetism depended on a metonymic relation between its opposing poles that explained how a compass appeared to move on its own. The ultimate source powering this magnetic motor remained metaphysical until 1600, when William Gilbert announced his discovery of geomagnetism and explained how it interacted with the compass to give it direction. Gilbert extended such magnetic interaction to all planets, claiming that it was the prime mover of the solar system. His influential conception of the universe as a kind of magnetic automaton constituted a major step toward secularizing the origin of movement and, in turn, the concepts of transformational motors and Romantic machines.

Whewell claimed that polarity was one of the main a priori ideas guiding scientific research and that it had materialized as a crucial object of study due to the identification of magnetic bipolarity. At the turn of the nineteenth century, such scientific interest in polarity motivated the research of proponents of *Naturphilosophie*, such as Oersted. His experimental proof of an intimate connection between electricity and magnetism confirmed a hypothesis that he had supported long before his discovery. During his formative years, Oersted had been a student of Schelling's *Naturphilosophie*. Schelling elaborated a scientific vision of the universe based on a theory of the unity of forces that depended on the bipolar structure of magnetism to explain other puzzling polarities, including mind and matter. The lodestone offered empirical proof that nature could be a self-contained autonomous whole that did not need an exterior spiritual power to set it into motion. Within Schelling's framework, magnetic bipolarity provided the most fundamental transformational motor to turn nature into a Romantic machine.

Elaborating on Jeremy Adler's study of Goethe's aesthetics of magnetism, I will discuss how the German poet also elevated magnetic bipolarity as a cosmic principle that helped apprehend and give form to nature's deepest mysteries. For Schelling and Goethe, magnetic bipolarity embodied the primordial expression of nature's paradox of unity in duality and consequently offered a privileged symbol for representing the metonymic relation that linked other polarities beyond that of mind and matter, including love and hate and conscious and nonconscious actions.

Schelling's and Goethe's *Naturphilosophie* mobilized a kind of analogical thinking that helped rehabilitate the scientific value of the old alchemical idiom of explaining the unknown by the unknown (*ignotum per ignocius*). Since Chaucer had mocked its obscurantism and Bacon had dismissed its usefulness, this alchemical logic had been used to discredit scientific works that seemed to rely too much on unfounded analogy. Yet a cursory look at the development of the concept of gravity shows that from Gilbert and Kepler to Faraday and Einstein, (electro-)magnetic analogies with little to no empirical foundation played a significant role. In poems invoking magnetic polarity, Goethe aspired to recover the lost unity of poetry and science through such alchemical logic. His analogy linking the mind and the compass yielded striking insight into the nature of the former and prefigured the psychoanalytic invention of the unconscious.

The polarity of life and death received one of its most well-known Romantic treatments in a key precursor of *L'Ève future*: Mary Shelley's *Frankenstein* (1818). Unlike electric interpretations of the secret to life in *Frankenstein*, I contend that magnetic polarity plays a more significant role in the novel to convey the elusive nature of life. The way the main characters converge and interact at the magnetic North Pole embodies a system of relations of contiguity that, like Schelling's and Goethe's, is anchored by magnetic polarity. The narrative structure of the novel functions like a compass, guiding the reader across the uncharted territory that metonymically links life and death.

In *L'Ève future*, Villiers revisits this territory with the added guidance of electromagnetic contiguity. Although magnetic bipolarity remains important to navigate this complex novel, it has been supplanted by a related yet different polarity where the two poles are now electricity and magnetism. To convey a notion of life that resists naïve mechanical appropriation, Villiers imagines an automaton powered by the metonymic power of induction. This new transformational motor manifests an epistemological shift that will become clearer after considering later electromagnetic analogies of life used by scientists such as Oliver Lodge and Jean Rostand.

Villiers suggests that phenomena associated with mourning are the ultimate secret ingredient for the creation of artificial life. His hybrid automaton allows him to represent and explore puzzling cognitive phenomena that magnetic somnambulists and hysteric patients had recently rendered manifest and that Joseph Breuer would later attempt to apprehend through a similar electromagnetic trope. In Villiers's and Breuer's interpretations of traumatic reminiscences as a kind of electromagnetic effect, the structure of the mind manifests the relation of contiguity between conscious and unconscious systems that marked the birth of psychoanalysis.

The Invention of Magnetic Polarity and Philosophy

The nautical compass originated in Asia and appeared in Europe at the end of the twelfth century.[11] By freeing navigation from its dependence on the sighting of landmarks and celestial bodies (and thus on clear weather conditions), the compass played a central role in the expansion of European trade and territories during the age of exploration. Due to its practical application in the compass, magnetism drew more and more scientific interest. The first breakthrough occurred in 1269 when, in *Epistola de magnete* (Letter on the Lodestone), Petrus Peregrinus (a.k.a. Pierre de Maricourt) invented the concept of magnetic polarity.

As with most natural philosophers before Copernicus, Peregrinus thought that the heavens revolved around the axis—or, in Latin, *polus*—passing through the North and South Poles of a motionless Earth. This vision of the cosmos came from the prevalent scholasticism of his time and its conception of the universe, deriving from Christian theology and Aristotelian physics. Within this framework, Peregrinus's famous contemporary Thomas Aquinas (1225–1274) could only explain the power of the lodestone by "its participation in heavenly virtue."[12] This explanation attributed the behavior of the compass to an exterior divine power, which connected with inert earthly things through intermediaries that could channel it. Such intermediaries included the Pole Star, because of its closeness to the North Pole, and imaginary magnetic mountains, which continued to appear on maps well into the sixteenth century.

In *Epistola de magnete*, Peregrinus states that the lodestone "bears in itself the likeness of the heavens" and, building upon the analogy with celestial poles, describes experiments designed to locate the lodestone's north and south poles. He calls the pole pointing north of the lodestone its *north pole*, and he records how the lodestone's north pole was respectively repelled and attracted by the north and south poles of another magnet. Interaction of magnetic poles caused attraction and repulsion.

As Peregrinus speculates on why opposite poles attract, he writes that magnetic polarity originates in active "virtues" transmitted from the poles of the "heavens." He also extends his newly formulated concept of magnetic polarity to the whole of nature, thereby inaugurating a philosophical tradition that, as we will see, would culminate with Schelling's *Naturphilosophie*. For Peregrinus, the cosmos consists of codependent "active" and "passive" agents: "The active agent requires a passive subject, not merely to be joined to it, but also to be united with it, so that the two make but one by nature."[13] An active heaven needs the passive Earth to maintain the unity of the cosmos. Even though the cosmos's unity is heavily dependent on an active heaven, it can only be sustained by an earthly, subaltern pole.

To justify this claim, Peregrinus follows the previous statement with the description of an experiment showing how a bisected lodestone produces two smaller lodestones. For him, these two new lodestones are not created equal. He called the one that ended with the north (or "active") pole of the original lodestone the "active agent" and the other one the "passive subject." Since the two magnets should have exhibited exactly the same characteristics, Peregrinus's interpretation betrays an erroneous valorization that derived from his scholastic framework and its hierarchical order of things. He claims that Earth's North Pole channels the virtue of the heavens and transmits its power to the North-seeking pole of the magnet. Yet as he demonstrated, only opposite poles attract, and therefore the north pole of the magnet can only point to Earth's South Pole. Such a misleading habit would persist until today. What we would commonly call a compass's magnetic north actually points toward the South Pole.

Peregrinus also observed that the two pieces of the original lodestone would naturally attract each other along the faces of the bisection. He interpreted it as a sign that "the active agent desires to become one with the passive subject because of the similarity that exists between them."[14] This "similarity" comes from the way they participate in magnetic power. Peregrinus describes a polarity structured by the duality of a dominating active pole of divine origin and its passive earthly counterpart, both of which, despite their fundamental difference, are magnetically connected. Like the link between divine and human represented since Plato by the magnetic chain, this connection manifests a metonymic relation between heavenly "virtues" and the earthly lodestone that informed Peregrinus's pioneering conception of magnetic polarity.

Peregrinus concludes *Epistola de magnete* with practical applications of the concept of magnetic polarity including designs for an improved compass and a "perpetual motion" machine. The latter consists of a wheel toothed with iron nails that interact with the poles of a lodestone: "When one of the teeth comes near the north pole and owing to the impetus of the wheel passes it, it then approaches the south pole from which it is rather driven away than attracted [. . .]. Therefore such a tooth would be constantly attracted and constantly repelled."[15] The compass and this supposed perpetual motion machine are early instances of magnetic automata. They both appear self-moving, yet their true motors are external. They become animate because they are part of a series of intermediaries that channel heavenly "virtues" to Earth. Following Peregrinus's scientific experiments with the lodestone, magnetic bipolar interaction offered a fresh way to explain how this divine motor worked.

Peregrinus was the main influence on William Gilbert, who, in *De Magnete* (1600), modernized the study of magnetism and electricity. An early convert to Copernican heliocentric cosmology, he disapproved of the scholastic conception of Earth as an inert body. Within this framework, the invisible power of the lodestone could only be explained through an otherworldly or external agent. During the sixteenth century, such an explanation failed to help navigators, who, for their own safety, had to account for the fact that a compass needle did not perfectly align with Earth's poles and pointed not up to the heavens or the North Star but down below the horizon. The identification of "the dip of the needle" in 1581 might have provided the most significant clue for Gilbert to search below the heavens for the real source of magnetic power.[16]

Gilbert argued that the earth was a giant lodestone and that its immaterial "sphere of virtue" controlled the behavior of the magnetic compass. He set up his experimental proof of geomagnetism with a lodestone replica of the earth (the *terrella*) and a miniature nautical compass (the *versorium*). The experiment worked. Moving like a boat around the *terrella*, the *versorium* exhibited magnetic variations and inclinations similar to what navigators had recorded during their voyages around the globe. Through this relatively simple analogy, Gilbert induced—and in turn discovered—that the earth generated its own magnetic field.

In *The New Organon* (1620), Francis Bacon lists the compass as one of the three main inventions responsible for the making of the modern world, praising Gilbert's inductive method as an important model with which to modernize the production of knowledge by recasting it on empirical grounds.[17] However, he also strongly criticizes Gilbert for extending the results of his experiments far beyond the domain of scientific investigation in his development of a "magnetic philosophy" based on the unfounded claim that the solar system moves due to a magnetic motor.

In 1543 Copernicus dealt a major blow to the scholastic conception of the earth as an inert body by arguing that the earth revolved around the sun while also rotating on its own axis. Identifying the agent responsible for the movement of the planets around the sun, Gilbert inferred that their motion was due to the earth's magnetic properties.[18] This theory informed his magnetic philosophy, which restored his planet's movement and dignity and which he associated with the holistic and alchemical thinking of "Hermes, Zoroaster, Orpheus." In lieu of an exterior agent or mysterious, unmoved mover, Earth changes position because it possesses a magnetic soul.

To describe how magnets—and by extension, planets—move in relation to each other, Gilbert rejected the popular designation of "attraction" (which to him implied a violent and one-sided exertion of force) and introduced the term *coition*.[19] Gilbert's idea of magnetic coition evokes a shared and harmonious interaction that starts with a preliminary, circular dance between two magnets floating on water and ends when they reach their natural resting positions. As the new motor powering the Copernican heliocentric universe, Gilbert's magnets explained how planets and stars move each other, and his magnetic philosophy sought to explore the laws of their interaction.

In typical Renaissance convention, Gilbert's magnetic philosophy revived ancient wisdom dating back to the pre-Socratics. In the first part of *De Anima*, as Aristotle surveys and criticizes various theories put forth by his predecessors concerning the soul, he writes, "Thales, too, to judge from what is recorded about him, seems to have held soul to be a motive force, since he said that the magnet has a soul in it because it moves the iron."[20] The magnet animates iron;

therefore, it has soul, and by extension, every animate or animating body like the magnet has soul. Although Thales's extant works are fragmentary, it appears that magnetism structures his animist thinking and its notions of soul and movement. On the verge of the seventeenth century, Gilbert essentially participates in this tradition of magnetic animism when he writes, "The Magnetic force is animate, or imitates the soul; in many respects it surpasses the human soul while that is united to an organic body."[21]

As seen with the aforementioned examples of the great chain of being and animal magnetism, such animist thinking depends on magnetism to make sense of other dynamic and elusive phenomena. For Gilbert, this analogical practice informed both the groundbreaking experimental proof of geomagnetism (with two models made from magnetic material representing the earth and the nautical compass) and the magnetic philosophy he derived from it. The discovery of geomagnetism led him to argue that the motor powering the Copernican solar system was magnetic, a speculation that Bacon dismissed as unscientific. Yet this bold claim represented a major step toward the secularization of the origin of movement and the rise of magnetic polarity as a transformational motor.

Alchemical Analogy: Explaining the Unknown by the Unknown

When Gilbert reviewed previous works on magnetism and electricity, he criticized the way his predecessors justified their claims by explaining "the unknown by the more unknown" in the manner of alchemists.[22] Gilbert eventually came under the same criticism. Although his analogy between the earth and the lodestone led to a major scientific breakthrough, it also proved misleading. He elaborated his magnetic philosophy based on the false supposition that Earth was a giant lodestone and wrongfully concluded that magnetism was the prime mover of the universe. A few years after Gilbert's death, Bacon compared him to the last alchemist (in Latin, *Chemicorum*) who could not refrain from inducing a universal principle out of a "narrow and unilluminating basis of a few experiments."[23]

More than two centuries earlier, Chaucer had written a satire about an alchemist, which subsequently inspired countless other parodies and critiques of a figure today associated with occultism and pseudoscience. Part of Chaucer's satire consisted of mocking the alchemical idiom *ignotum per ignocius*—explaining the unknown by the more unknown. In the subnarrative at the end of *The Canon's Yeoman's Tale*, a student asks his master, the alchemist "Plato," about the secret of the philosopher's stone. The alchemist replies that the stone is called "Titanos," prompting the student to ask what that stone is. Plato replies, "*ignotum per ignocius*"—that "Magnasia is the same."[24]

Bacon invokes the alchemist's deliberate use of obscure language to transmit secret knowledge because it also conceals errors.[25] Gilbert's experiments led to the discovery of geomagnetism, but he pushed the analogy between magnetism, the earth, and gravity too far. Based on the comparison between the invisible power of magnetism with other elusive phenomena, magnetic analogies always rely to some degree on the alchemical logic of *ignotum per ignocius*. Such analogies are particularly transparent in magnetic animism and its conceptions of the soul and of movement. Logicians argue that this alchemical logic begs the question and/or fails to elucidate.

Despite this fallacy, Gilbert's magnetic animism had a tremendous influence on the seventeenth-century natural philosophers who paved the way for modern physics. By 1603 Johannes Kepler could affirm that he would "demonstrate all the motions of the planets" thanks to the physical model for gravitation that he derived from Gilbert's theory of magnetic cosmology. In contrast to his holistic notions of magnetic animism and coition, Kepler developed a more mechanical conception of the universe. He claimed that his aim was "to show the celestial machine not to be [a] divine organism but rather to be a clockwork [. . .], all the [. . .] movements are carried out by a single, [. . .] magnetic force, as in [. . .] a clockwork all motions by a single weight."[26] Kepler considered the universe a machine and magnetic force to be its prime motor. This mechanical analogy helped him conceive of a more predictable model of causal interaction, from which he derived his laws of planetary motion.[27] In 1685 Kepler's achievements in magnetic mechanism played a central role for Newton when, in the *Principia*, he unveiled the classical laws of mechanics.

Newton focused on gravitational attraction and left it to his followers to explain the mysterious behavior of magnetic attraction. Throughout the eighteenth century, the Cartesian model of magnetism usually prevailed.[28] For Descartes, Gilbert's magnetic philosophy remained too entrenched in magical traditions of soul-like immaterial forces. In 1644 Descartes published in *Principia Philosophiæ* a materialist explanation of magnetism to expunge its occult interpretation from his own clockwork universe. To illustrate his claim, Descartes provided a woodcut reproduction that features, in Gilbertian fashion, a lodestone shaped into a *terrella* and surrounded by five other satellites (fig. 10).[29] Magnetic particles, or *effluvia*, travel around the lodestone from its

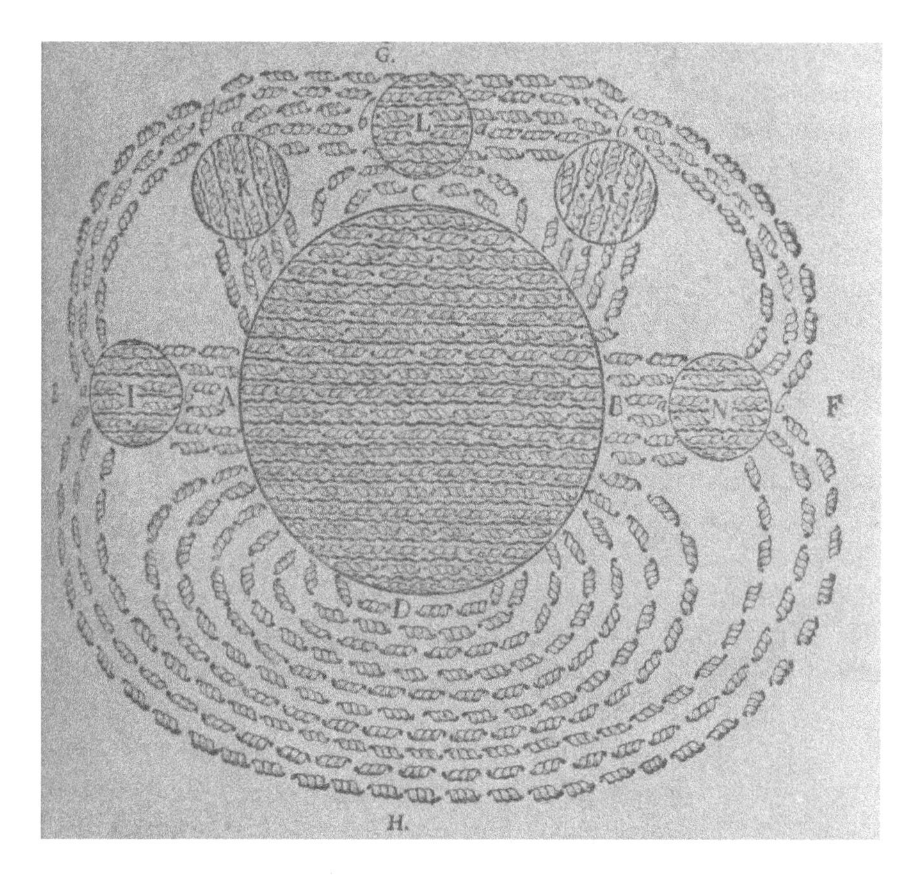

Fig. 10 Descartes's magnetic effluvia diagram. René Descartes, *Principia Philosophiæ* (Amsterdam: Apud Ludovicum Elzevirium, 1644), 273. Library of Congress, Washington, DC.

north pole to its south pole and inside through the lodestone's pores. They come outside again through its north pole due to hollow channels connecting the two poles. Descartes conceives of the magnetic particles as screw-threaded to represent their one-way flow and how they attract distant objects.

Descartes's magnetic mechanism functions due to the medium provided by these invisible yet material effluvia, which are responsible for motion. He also relied on a similar materialist model to explain gravity. In the latter, vortices operating in a vacuumless space caused a gravitational pull. This contrasted with Newton's concept of universal gravitation. Newton translated the attracting power of gravitation into a quantifiable force that he could easily calculate by following a set of mathematical equations. However, he never really dared to consider the medium responsible for the interaction between distant bodies as material, suggesting in turn a vacuum between them.

Like Gilbert's magnetic philosophy before it, Newton's theory of mechanics came under attack from Cartesians, who accused him of lapsing into occultism by failing to posit a tangible medium or physical agent to explain gravity's action at a distance. In one of his 1734 philosophical letters dedicated to Newton, Voltaire illustrated the cleavage between Cartesians and Newtonians by writing that the trip from Paris to London was like leaving a world of plenitude for an empty one. Newton handed down to his successors the task of identifying the medium—if there was one—responsible for universal gravitation. To substantiate the elusive power of gravity and magnetism, natural philosophers throughout the eighteenth century offered system after system, trying to fill in the spatial disconnect left by Newton.[30]

A close friend of Newton's, Edmond Halley (1656–1742) was one of the first major figures to combine Cartesian effluvia with Newtonian mechanics in his research on magnetic variation. Halley became famous when he calculated the orbit of the comet that would subsequently receive his name, and he correctly forecasted its next return with the help of Newton's mechanics. Halley did not live to see the spectacular confirmation of his calculations in 1758 or how they contributed to converting many of the remaining skeptics to Newton's theory of universal gravitation.

When Newton's *Principia* appeared in 1685, Halley publicized it by ridiculing previous Cartesian conceptions of gravity for being "incomprehensible." He also directed his attack toward the erroneous alchemical analogy of gravity and magnetism: "Some think to illustrate this *Descent* of *Heavy Bodies*, by comparing it with the Vertue of the *Loadstone*; but setting aside the difference there is in the manner of their *Attractions*, the *Loadstone* drawing only *in* and *about* its *Poles*, and the *Earth* near equally in all of its *Surface*, this Comparison avails no more than to explain *ignotum per aeque ignotum*."[31] Halley attacks the natural philosophers who, like Kepler after Gilbert, confuse gravity with magnetism. As with Chaucer and Bacon, he focuses his critique on the alchemical analogy between magnetism and gravity and how it comes down to explaining *ignotum per aeque ignotum*. For Halley, magnetism does not work as a physical model for gravitation, especially since both manifest divergent forms of "Attractions." Schematically, gravity stems from one center exerting a constant and linear force everywhere on the surface of a spherical object like Earth, whereas the magnet's "Attraction" derives from two centers, or "Poles," which produce a more complex sphere of influence. Later in his life, Halley himself tried to reverse the alchemical analogy he had criticized by comparing magnetism to gravity. However, mathematical laws similar to what Newton had devised for gravitational force did not comply with the calculation of magnetic force.

In 1634 Gilbert's magnetic philosophy suffered a major setback with the discovery that magnetic variation changes over time. Gilbert's simple Earth-*terrella* model could not account for the changing position of the planet's magnetic poles. Halley, who continued to believe that Earth was a giant magnet, came out with a new hypothesis to account for Earth's shifting magnetic poles, which was to remain influential well into the nineteenth century. Inspired by Saturn and its rings, he argued that Earth consisted of at least two rotating parts made from magnetic matter: a hollow shell and a globe within it. Halley concluded that magnetic variation resulted from a discrepancy between their respective speeds of rotation. Since life thrives on the outer part of the shell, Halley inferred that other "creatures" must live on the surface of the inner globe.

Long before Jules Verne, Halley's sheltered magnetic core inspired stories about extraordinary voyages to the center of the earth and other hidden worlds. One of the earliest and most famous examples is *Gulliver's Travels* (1726), in which Jonathan Swift satirized the idea of a secret magnetic world with the floating magnetic island of Laputa.[32] Halley pushed his speculation even further when he claimed that life was possible in Earth's inner magnetic world due to a subterranean source of light. In an article from 1716 on the aurora borealis (northern lights), Halley attempted to solve two mysteries with one theory. In a Cartesian fashion, he invoked "Magnetical Effluvia" slipping out of the earth's pores to explain the aurora borealis, while at the same time claiming that this magnetic light must somehow be produced to illuminate the inner world. Unable to discover how the magnetic effluvia generated light, he evoked an "Affinity" existing among various effluvia and drew a parallel with the example of the electric spark.[33]

Throughout the eighteenth century, alchemical analogies and apparent affinities between gravity, magnetism, electricity, and light made the elucidation of these phenomena often interchangeable, since materialist euphemisms such as *effluvia* or *fluid* could be attributed to just about anything that could be felt but whose cause eluded the eye. They also suggested a common unity with these occult forces that could explain why magnetism and electricity seem to light up just as easily. In *De Magnete*, Gilbert had created a radical division between magnetism and electricity by considering the former immaterial and the latter material. Even though many did not believe in the affinity of electricity and magnetism, fluid theories such as Halley's made the link between magnetism and electricity plausible for some, yet it was still in need of experimental proof.

In a 1773 letter entitled "On the Analogy Between Magnetism and Electricity," Benjamin Franklin writes, "As to the magnetism, which seems produced by electricity, my real opinion is, that these two powers of nature have no affinity with each other, and that the apparent production of magnetism is purely accidental."[34] The belief that electricity and magnetism were unrelated particularly informed French and British natural philosophy. For instance, published the year before the discovery of electromagnetism, an 1819 article on magnetism from the *Edinburgh Encyclopedia* stated, "It is proved that every piece of iron which has suffered any friction becomes magnetic, and that an electrical discharge, acting like a blow, develops magnetism in iron wires through which it is made to pass. From the same cause, lightning produces a similar effect upon the mariner's needle, and sometimes even reverses its poles."[35] The puzzling affinity between electricity and magnetism, manifested in a familiar phenomenon like the effect of lightning on a compass needle, disappears in the mechanical explanation. These magnetic effects do not occur through some kind of intimate relation with electricity but through a mechanical action such as "friction" or an electrical "blow."

A year after this article appeared, Oersted announced that he had discovered a metonymic link between electricity and magnetism and, as previously discussed, triggered a series of experimental and conceptual breakthroughs that would legitimize alternative modes of scientific thinking that could no longer solely rely on such mechanical explanation. On some level, electromagnetism confirmed the validity of ancient alchemical analogies among different natural forces as a motor for scientific exploration. The "happiest thought" that helped Einstein figure out how to include gravity within his general theory of relativity occurred due to a thought experiment that staged an analogy between electromagnetic induction and gravity. This analogy was participating in a long tradition that, from Gilbert's and Kepler's magnetic interpretations of gravity to Faraday's use of magnetic lines of force to reconceptualize gravitational attraction,[36] yielded epoch-making discoveries about the nature of the universe.

These (electro-)magnetic analogies have invariably relied to some degree on the alchemical logic of explaining the unknown by the unknown. An influential intellectual movement predating electromagnetism that attempted to rehabilitate the scientific value of alchemical logic was *Naturphilosophie*.[37] Unlike most of his French- and English-speaking contemporaries, Oersted actively sought to prove the affinity between electricity and magnetism because he had been an enthusiastic student of *Naturphilosophie*. I will focus on the critical role magnetic polarity played for two of the most prominent figures associated with it, Friedrich Wilhelm Joseph von Schelling (1775–1854) and Johann Wolfgang von Goethe (1749–1832). Instead of an outside impetus, in their works magnetic bipolarity functions as a transformational motor that moves on its own due to the difference marking the relation of contiguity of its poles. Magnetic bipolarity helped both men conceive of the universe as a self-contained Romantic machine whose deepest mysteries could only be intelligible through analogical means that went beyond mere mechanical thinking.

The Logic of the Compass

The most prominent theorist of *Naturphilosophie*, Friedrich Wilhelm Joseph von Schelling (1775–1854), opposed the prevalent mechanical cosmology of the end of the eighteenth century with a philosophy modeled in great part after the polarity (*Polarität*) of the magnet. He conceived of nature as a self-contained organism, which could not be understood solely as a classical machine but rather had to be fathomed through correspondences that analogically linked nature's innermost secrets with its constituents—more specifically, with man, magnetism, electricity, and chemistry.[38] Although different, these exemplary constituents shared a secret unity through a system of metonymic and metaphoric relations.

The notion of nature as a self-contained organism faced the problem of differentiation.[39] With no outside to nature, its scientific investigation could only happen from inside. Nature must then generate difference from within, enabling it through self-differentiation to become its own object of study. It follows that nature is both self-identical and plural. To explain the apparent paradox of this self-identical plurality, or "identity in duplicity and duplicity in identity," Schelling relies on the lodestone, the most elemental example of polarity: "In magnetism, we see in the whole of *nonorganic* nature that which is really the character of Nature as a whole—namely, identity in duplicity and duplicity in identity (which, said otherwise, is the expression of polarity). It should be said that every magnet is a symbol of the whole of Nature."[40] Nature expresses its mysterious principle of plurality in identity in this stone, where opposite poles share a relation on contiguity that is indivisible.

As a differentiated whole, the lodestone structures the conceptual framework of Schelling's *Naturphilosophie* and its attempt to reconcile other polar opposites such as matter and spirit. Schelling conceived his philosophical system as a critique of classical materialistic and mechanical notions that considered "spirit" as something exterior to nature. Whereas magnetism provided an external spiritual impetus to trigger the clockwork of Kepler's mechanical universe, for Schelling it manifested a fundamental bipolarity that set everything in motion from the inside. It also rendered other polar opposites intelligible, since they were all expressions of the same fundamental principle of nature.

The poetry of Goethe illustrates this complex system of metonymic and metaphoric relations, which Schelling extracted from magnetic bipolarity. Inspired by *Naturphilosophie*, Goethe had planned to follow up the poem version of *Metamorphose der Pflanzen* (1798) with a similar piece on magnetism.[41] As Jeremy Adler has shown, it is in a series of couplets from *Gott, Gemüt und Welt* (1815) that Goethe relied most transparently on magnetic polarity to give form and content to his verses. For instance, in the following verses,

> Magnetes Geheimnis, erkläre mir das!
> Kein Größer Geheimnis, als Liebe und Haß
> (Magnetic secret, explain me that!
> No bigger secret, than love and hate)

Goethe draws an analogy between the secret (*Geheimnis*) of magnetic polarity and that of love and hate. He does so by bringing together another set of polar opposites: folk wisdom and scientific thinking. The rhyme of *das* on *Haß* imitates the coarseness of peasant talk while revealing a fundamental truth about conflicting forces in nature.[42]

In this couplet, Goethe indulges in popular art as part of the larger Romantic program that he shared with Schelling and that consisted of recovering the lost unity of science and poetry. When it comes down to apprehending the deepest secrets of nature, folk wisdom could be just as illuminating as Enlightenment science ("*erkläre mir das!*"). Goethe's rustic couplet explores one secret in terms of another. It works through an alchemical analogy that explores the unknown by the unknown. The magnetic analogy rehabilitates the poetic approach to knowledge by implying a resemblance between the secret metonymic polarities of magnetism and of love and hate. The elusive secret of magnetic bipolarity expresses a metonymic link that connects everything in nature and allows for what appears to be unrelated phenomena to be mutually illuminating.

This alchemical method not only helps reconcile poetics and science but also implies an ethical approach to knowledge production contained in two other couplets concerning the autonomy of the self:

Was will die Nadel nach Norden gekehrt';
Sich Selbst zu finden, es ist ihr verwehrt
(Why does the needle turn to the North?
It will never find itself.)

Soll dein Kompaß dich richtig leiten,
Hüte dich vor Magnetstein', die dich begleiten.
(To keep your compass a proper guide,
Beware of the irons that cling to your side.)[43]

These verses expand on the idea of a "moral compass" by revealing that the self is also under the sway of something beyond itself. Like a compass, the self receives guidance from an outside influence that points it toward a cardinal point to navigate the unknown. Yet the actual reason for the behavior of pointing refuses to be told ("*Sich Selbst zu finden, es ist ihr verwehrt*"). As long as no other magnets interfere with it ("*Hüte dich vor Magnetstein', die dich begleiten*"), the self is swept (*gekehrt*) by a power that forces its poles to align with those of nature.

This power beyond the self is not simply outside of it. For geomagnetism to impose its direction, the needle of the compass must also be magnetic. All magnets possess opposite poles, which, as Peregrinus discovered, are responsible for their mutual attraction and repulsion. Magnetic bipolarity must already be present in what it can influence. The reason the needle turns north lies outside itself, beyond its grasp, but also within itself, in the way it is structured so as to be receptive to the direction of geomagnetism.

The analogy of the compass alludes to a state of being that is under the influence of an outside that is also already inside. This influence secretly operates through an outside-inside (or, correspondingly, inside-outside) logic upon which, Schelling would argue, rests our ability to fathom nature's deepest secrets. In Goethe's couplets, this logic of the compass also emphasizes that the self is remotely guided by something beyond yet within itself. The self is a Romantic machine powered by a secret bipolarity expressed by magnetism, which—long before Joseph Breuer and Freud—reveals itself to be under the sway of conscious and nonconscious systems sharing a relation of contiguity.

The Polar Secret of Life

During the nineteenth century, the secret of geomagnetism and magnetic variation continued to elude scientists, who attempted to solve its mystery by organizing dangerous expeditions to remote and uncharted polar regions. Participating in one of the biggest international scientific efforts of the time,[44] these real-life adventures inspired literary authors to mobilize the mystery of geomagnetism as a motor for generating narratives that, like Schelling's and Goethe's *Naturphilosophie*, relied on this macrocosmic instance of magnetic bipolarity to explore other secrets. The most famous example is Mary Shelley's *Frankenstein* (1818). In this novel, all the mysteries associated with the creation of life converge in the Arctic region, where, through the interactions of the characters, they begin to share a relation of contiguity. This convergence of secrets manifests a logic of the compass where alchemical analogies centered on the magnetic North Pole give form to a narrative structured around the polarity of life and death.

Shelley sets the magnetic stage of *Frankenstein* with the first letter Walton sends to his sister, describing his upcoming expedition to the North Pole and the scientific expectation that such an adventure would inspire in nineteenth-century explorers: "What may not be expected in a country of eternal light? I may there discover the wondrous power which attracts the needle [. . .]. You cannot contest the inestimable benefit which I shall confer on all mankind to the last generation, [. . .] by ascertaining the secret of the magnet, which, if at all possible, can only be effected by an undertaking such as mine."[45] Although he is equipped with a compass pointing to the source of "the secret of the magnet," Walton will only be led to encounters with other secrets.

Once in the desolated regions of the Arctic Circle, Walton crosses the path of the other best-kept secret in *Frankenstein*: the secret of life. When his ship, enclosed in ice, is nearly brought to a stop, he perceives a strange humanoid form on a sledge drawn by dogs and moving northward. The next day, he rescues Doctor Frankenstein from a drifting parcel of ice. Frankenstein has been running after his "creature"—a makeshift humanoid body to which he had given life. Inspired by alchemists like Paracelsus and the latest progress of natural philosophers in chemistry, electricity, and physiology, Frankenstein had studied "the change from life to death, and death to life" and eventually uncovered an "astonishing secret" when he discovered "the cause of generation and life." Convinced that the polarity of life and death revolved around a surmountable difference, he had then taken it upon himself to reanimate a body made up of dead body parts.[46]

Walton has finally met a companion who appears to share his own ambition for discovery. Despite Walton's eagerness to learn about his new friend's finding, Frankenstein does not share the secret of life. For Frankenstein, this secret has come to symbolize the scientific hubris that has brought him only trouble and despair. He describes the fateful night he gave life in elliptical terms: "I collected the instruments of life around me, that I might infuse a spark of being into the lifeless thing that lay at my feet. [. . .] I saw the dull yellow eye of the creature open."[47] Contrary to Hollywood's subsequent electric interpretations, the expression "spark of being" maintains a more ambiguous representation of the "principle of life." According to Marilyn Butler, *Frankenstein* reflects—without taking a clear side—the highly publicized vitalist debate that took place in London between 1814 and 1819.[48] Proponents of "spiritualized vitalism" came under materialist attacks for explaining the principle of life in terms of "illusionary analogies" with other forces such as gravity, chemical affinity, electricity, and galvanism. Materialism, for its part, focused only on "the organization and function of living bodies" and in turn sidestepped the difficulties of explaining life itself. Frankenstein's interest in alchemy and his coldblooded bodysnatching respectively incarnate the two extreme sides of this debate. Even though his vague account remains inconclusive, for Butler the description of the creature's creation depends more on spiritualized vitalism than materialism, since Frankenstein's "instruments of life" bring to mind the voltaic battery.[49]

Considering Walton's first letter and the central role the symbol of magnetism played in Schelling and Goethe—and more broadly, in the Romantic movement—I would argue that Shelley's analogy with magnetic polarity is important for navigating the narrative of *Frankenstein*. Shelley's novel conveys at least one certainty: secrets attract. The North Pole holds for Walton the secret of the magnet, but it also draws in the creature, who is the living proof of Frankenstein's secret. Besides revengefully luring Frankenstein into the harshest of terrains, the creature appears destined to reach the North Pole. When Frankenstein passes away, Walton finally comes upon the creature, after it had surreptitiously entered his boat to mourn its maker. Before vanishing, it tells Walton, "[I] shall seek the most northern extremity of the globe; I shall collect my funeral pile, and consume to ashes this miserable frame, that its remains may afford no light to any curious and unhallowed wretch, who would create such another as I have been."[50] The secret of the polarity of life and death dies with Frankenstein and the self-destruction of his creature. *Frankenstein* ends on the edge of a magnetic pole, where the secrets of life and death—and, arguably, electricity—all converge.

This dramatic denouement implies that, like Schelling's and Goethe's logic of the compass, the secret of magnetic polarity is the anchor point of all these different secrets. It provides a center on a barren map where the narrative is set into motion and where all the secrets of the novel establish relations. These relations are not simply metaphoric like in the analogy of life and the electrical "spark"; they are also deeply informed by the metonymic link expressed by magnetic bipolarity, which Shelley mobilizes to convey the connection responsible for the polarity of life and death.

Twenty years later, Poe would write a narrative also terminating at a pole. *The Narrative of Arthur Gordon Pym* (1838) is a fictional travelogue that recounts the eponymous character's adventures to the South Pole. Poe drew inspiration from Jeremiah N. Reynolds (1799–1858), who, during the 1820s and 1830s, petitioned the U.S. Congress to support his Antarctic expedition to verify if the North and South Poles were actually "holes."[51] Halley's 1716 claim that Earth was hollow and produced magnetic effluvia illuminating an inner world and producing the northern lights led to various speculations concerning the poles, which included the popular theory that they were two large holes. This theory also provided an explanation for the motor propelling ocean currents, since water would fall into the holes and in turn circulate from one end of the planet to the other.

In the last few entries of his travelogue, Pym describes how he is on a boat approaching the South Pole and sees "a limitless cataract, rolling silently into the sea from some immense and far-distant rampart in the heaven."[52] This silent and undulating veil flowing from the sky to the sea is most likely the southern lights (aurora australis), the southern hemisphere's counterpart to the northern lights. The current is pulling Pym and his shipmate faster and faster toward the South Pole and "into the embraces of the cataract, where a chasm threw itself open to receive [them]" and where they perceive an enigmatic figure "of the perfect whiteness of the snow." As Pym is about to penetrate the secret of geomagnetic polarity, silence and whiteness dominate his account. Throughout his journey at sea, Pym went through many colorful adventures, but as he is about to apprehend the secret of the poles, his narrative abruptly ends.

Frankenstein and *The Narrative of Arthur Gordon Pym* are textual Romantic machines that rely on the secret of magnetic bipolarity to drive narrative action. This bipolarity expresses a relation of contiguity that structures their stories and gives their secrets a certain coherence that comes from explaining the unknown by the unknown. Here the alchemical analogy has not yet registered the implications of electromagnetism, which began to emerge a few years after the publication of *Frankenstein* and which Poe started to exploit more systematically in the later mesmeric tales examined in chapter 1. Although these polar narratives and Poe's mesmeric tales convey the secret polarity of life and death by emphasizing its relation of contiguity, the physical model informing this metonymic relation has changed: Poe's interest in animal electromagnetism brought to the fore another transformational motor that derived its movement not from magnetic bipolarity but from a new kind of polarity where the interrelated poles were now electricity and magnetism. This change not only expressed recent changes in the physical understanding of nature, but it also brought more empirical support for exploring relations of contiguity among what appeared to be, like life and death, radically different.

Later theories that relied on analogies with electromagnetism to penetrate the secret of life substantiate this last claim. Before turning to Villiers's automaton and its electromagnetic interpretation of artificial life, I will clarify some of the epistemological implications associated with the shift from magnetic to electromagnetic contiguity by addressing a similar analogy of life proposed later by Maxwell's close follower and pioneer of wireless telecommunication, Oliver Lodge (1851–1940). He and French biologist Jean Rostand (1894–1977) relied on electromagnetic interaction to make sense of the polarity of life and matter and to avoid the conceptual pitfalls that had marked earlier debates between the proponents of vitalist and materialist conceptions of life.

Electromagnetism and the Polarity of Life and Matter

In *La vie et ses problèmes* (Life and its problems, 1938), Rostand writes, "They say a living being is always born of a being similar to itself. But before the discovery of electrical magnetization, one could also have said that every magnet originated from a pre-existing magnet."[53] Although Rostand does not dwell on this analogy, it clearly alludes to an epistemological change influenced by electromagnetism. Before Oersted, Ampère, and Faraday, magnetization and the way a bisected magnet turned into two magnets conveyed a repetitive action whereby same gave birth to same. A new type of reproductive process became apparent with the recognition that an electric current automatically gave rise to a magnetic field. For Rostand, this electromagnetic phenomenon provided a new physical model for thinking about a reproductive process marked by difference and, consequently, a metonymic relation between the begetter and the begotten.

The source of Rostand's electromagnetic analogy of life is Lodge's *Life and Matter* (1905), which provides more details concerning what is epistemologically at stake in this metonymic shift. Lodge used this analogy as an alternative to the "cheap monism" of German biologist Ernst Haeckel (1834–1919).[54] Haeckel had claimed that life was just a by-product of "matter." For Lodge, this theory faces at least two major challenges: (1) what "matter" really is remains unclear, and (2) by what process inert matter turns into life continues to be a mystery. Lodge claims that Haeckel can only solve this mystery in terms of "spontaneous generation," a vague idea that begs the question of exactly what in matter gives life to life. Lodge follows his critique of Haeckel with his own theory that life is a "permanent entity" independent of—but rendered temporally manifest by—matter. Given the state of knowledge about the nature of life at the turn of the century, he admits that he can only defend his claim through analogy and deems electromagnetism the best model available.[55]

As in Rostand's analogy, Lodge divides the history of magnetism between a before and an after electromagnetism. Before, scientists understood the production of magnetism as deriving from a source of magnetism. After Faraday and Maxwell, that view changed: "But within the nineteenth century, a fresh process of magnetisation has been discovered, and this new or electrical process is no longer obviously dependent on the existence of antecedent magnetism, [. . .] electricity was set in motion, constituting what is called an electric current, magnetic lines of force instantly sprang into being, without the presence of any steel or iron; [. . .] These electrically generated lines of force are similar to those previously known, but they need no matter to sustain them. They need matter to display them, but they themselves exist equally well in perfect vacuum." The magnetic field generated by an electric current proved that magnetism could also result from a manifestly different source and that it did not depend on matter to exist. Magnetism preexists its material incarnation. Correspondingly, life has always been a kind of immaterial "infinite reservoir," which has been rendered manifest through entities with which it does not seem to have anything in common:

> Magnetic behaviour exhibits a very fair analogy to some aspects of that still more mysterious entity which we called "life"; and if anyone should assert that all magnetism was pre-existent in some ethereal condition; that it would never go out of essential existence; but that it could be brought into relation with the world of matter by certain acts,—that while there it could operate in a certain way, controlling the motion of bodies, interacting with forms of energy, producing sundry effects for a time, and then disappearing from our ken to the immaterial region whence it came;—he would be saying what no physicist would think it worth while to object to,—what many, indeed, might agree with.[56]

Lodge answers the question about how the material and the immaterial interact without recourse to vague monist notions such as "spontaneous generation." He does so by relying on a metonymic interpretation of electromagnetic phenomena where electricity, magnetism, and matter share a relation of contiguity.

According to Rostand, before the advent of electromagnetism, magnetism could offer an analogy based on a metaphoric relation where same would beget same. Yet the central role magnetic bipolarity played in Romantic thought prefigured the metonymic shift expressed by Lodge's and Rostand's electromagnetic analogies of life. Inspired by Schelling's *Naturphilosophie*, Goethe drew from the relation of contiguity expressed by magnetic bipolarity to fathom how nature could contain radically conflicting elements. Yet although magnetism symbolizes other phenomena through its resemblance of their bipolar structures, it can only do so by universalizing the relation of contiguity that makes its poles indivisible. Such logic of the compass informs *Frankenstein* when the secrets of magnetism and life meet at the North Pole to establish a relation based on the contiguity of magnetic poles and of life and death. Historically situated between *Frankenstein* and Lodge's theory of life, Villiers's *L'Ève future* builds upon the metonymic reasoning associated with the Romantic appropriation of magnetic bipolarity, while it also manifests a transition from magnetic to electromagnetic contiguity that has registered some of the latter's epistemological implications.

From its invention in the Middle Ages to its symbolic appropriation during the nineteenth century, magnetic polarity guided physical and conceptual explorations that drew their impetus from the metonymic interaction of its poles. In Schelling's and Goethe's *Naturphilosophie* and Shelley's *Frankenstein*, magnetic polarity provided a transformational motor that turned the universe, the living, and the self into self-moving Romantic machines. The compass (but also so-called perpetual moving machines like Peregrinus's) inspired and provided empirical support for such conceptual automata.

In his previously mentioned seventeenth-century writings on the all-encompassing power of magnetism, Kircher discusses various other examples of magnetic automata that he collected, constructed, and displayed in his museum at the Jesuit college in Rome. He allegedly built a magnetic clock and humanoid automata, which participated in biblical scenes that included Jonas getting swallowed by the whale and Jesus walking on water and rescuing Peter from drowning. For the latter, Kircher explains, "A strong magnet must be placed in Peter's breast, and the hands of Christ, stretched out to come to the rescue, or any part of his toga turned toward Peter should be made of excellent steel, and you will have everything required to exhibit the story [. . .]. This will happen with greater artifice if the statue of Christ is flexible in its middle, for in this way it will bend itself, to the vast admiration and piety of the spectators."[57] Within Kircher's vision of the magnetic chain of being, such self-moving machines not only entertained spectators but also provided a legitimate reflection of how the universe and the divine worked.

A direct heir of Kircher's magnetic automata, Goethe's alchemical logic of the compass, and *Frankenstein*, Villiers's *L'Ève future* depicts in great detail the various elements that must go into the construction of an automaton that could replace a human being. To convey the secret of artificial life, Villiers relies on the new transformational motor of electromagnetism, its implementation in animal electromagnetism, and Edison's technologies. I will concentrate on the representation of animal electromagnetism throughout this complex novel to show how it allows Villiers to continue the Romantic critique of the mechanical interpretation of life by shifting the emphasis from magnetic to electromagnetic contiguity.

L'Ève future begins with an atmospheric description of Edison's laboratory reminiscent of an alchemist's den crowded with its typical paraphernalia of mysterious electrical machines and "enormous magnets."[58] In the scene, Edison is lamenting that his phonograph came too late and that he was not able to record the famous voices and sounds of history. He wonders why the "Word Made Flesh" missed out on the opportunity to speak into the horn of his phonograph to have his actual voice recorded:

> It's apparent, he resumed, that the Word Made Flesh paid little attention to the exterior and sensible parts either of writing or of speech. He wrote on only one occasion, and then on the ground. No doubt He valued, in the speaking of a word [*dans la vibration du mot*], only the indefinable *beyondness* with which personal magnetism inspired by faith can fill a word the moment one pronounces it. [. . .] Still, the fact remains, He allowed men only to print his testament, not to put it on the phonograph. Otherwise, instead of saying, "Read the Holy Scriptures," we would be saying, "Listen to the Sacred Vibrations!"[59]

Edison speculates why the "Word Made Flesh" never bothered to leave direct and lasting traces of his discourse in written or phonographic forms. The latter inscribes the vibrations produced by sound on the soft material of a rotating cylinder. Now that his invention showed that the spoken word could easily be recorded through a process of inscription akin to writing, allowing for the original utterance to be replayed with fidelity, Edison can poke fun at Christian phonocentrism.[60] The "Word Made Flesh" did not foresee that the living word would become reproducible or that its "magnetism inspired by faith" would eventually be mechanically duplicated. After the invention of the phonograph, the magnetic power of the sacred became recordable vibrations.[61]

Edison's ruminations are interrupted when he receives the visit of Lord Ewald, an old friend who had once saved him from penury. Lord Ewald confesses to Edison that he has decided to end his life for having fallen in love with a "bourgeois Goddess."[62] Although Miss Alicia Clary possesses the "ideal" body of the Venus de Milo, she has the mind of a philistine. Lord Ewald explains how her ways are terminally linked to the most pedestrian tastes, using a simple magnetic trope: "There's a deep but hidden correspondence between certain people and these non-topics [*ces choses inférieures*], a kind of reciprocal attraction or instinctive magnetism [*cette naturelle tendance, cet aimant réciproque*] that draws them together. One calls to the other, they are attracted, they draw together, and mingle."[63] This analogy based on resemblance emphasizes how the similar tends to attract the similar.

Exasperated by the "non-correspondence" between her physique and her spirit but unable to figure out why he is still attached to her, Lord Ewald shifts from the metaphoric to the metonymic aspect of magnetism to express his helplessness in terms of bipolarity: "So it is that this mistress, an animated dualism who repels and attracts me simultaneously, holds me to her by that very process, as the two poles of this magnet attract by their contradictory impulses a bit of steel."[64] The secret of polarity, or what Schelling called "duplicity in identity," is the source of her power of attraction. The pronounced dualism that Miss Alicia Clary embodies has afflicted Lord Ewald with a kind of bipolar disorder.

Villiers peppered *L'Ève future* with familiar Romantic polar images. Edison inflates the mystery of the North and South Poles by implying that the great flood was due to an "oscillation at the poles," before boasting a few pages later that his telephone can transmit a conversation all the way from the North Pole.[65] When he evokes the existence of a medium responsible for second sight, he compares it to the force that makes a magnetized needle point to the North Pole. And despite the low latitude of Menlo Park, New Jersey, Villiers cannot refrain from draping the sky during the climax of the novel with the aurora borealis.[66]

Yet magnetic polarity does not provide the main conceptual guide for navigating the complex network of ideas that Villiers brings together in the novel. According to Edison, Lord Ewald's bipolar disorder derives from a bad case of outmoded idealism that cannot adapt to the new electromagnetic law of modern love: "At bottom, modern love (if it is not, as contemporary physiology would have it, simply a matter of mucous membranes) looks to a physical scientist like a mere matter of equilibrium between a magnet and the object it attracts [*et une électricité*]. Thus, consciousness, without being wholly alien to the phenomenon, is perhaps indispensable only in one of the two poles. It's an axiom confirmed by a thousand experiments every day, notably those involving suggestion."[67] Magnetism can induce an electric current in a conductor, and vice versa. Modern love consists of achieving a certain "equilibrium" in this mutual induction. Yet the reciprocal convertibility of these two forces also suggests that only one is necessary to generate the other. A modern couple could thus potentially thrive on one individual as the magnetic "consciousness" generating electricity even in the most inane partner. The invisible medium of love, desire, or "suggestion" operates through a kind of electromagnetic induction.

Edison has found the antidote to the pangs of love, which have led so many young idealists to suicide. He undertook to find it after another of his esteemed friends, Mr. Edward Anderson, gave up a happy marriage, and eventually his life, for a frivolous liaison. As another instance of "the luminous principle of the attraction of opposites,"[68] Edison demonstrates to Lord Ewald how the respectable Mr. Anderson became the victim of his own overvalorization of an unattractive and conniving lover. Like Mr. Anderson, Lord Ewald is caught between his love for an illusion and the unfortunate reality onto which it is projected. Edison proposes to cure such Romantic bipolar disorder by having the pole of illusion completely overtake the depressing pole of reality. This shift could be achieved through the construction of an "Android"—a "magneto-electric entity" that will be made after Miss Alicia Clary's image but omit her repulsive bourgeois personality.[69]

Edison has already built a prototype called Hadaly and proceeds to show Lord Ewald how his remedy for heartbreak functions. Electromagnetism powers this new Eve. Just as motion is generated in the stylus of Morse's telegraph, Hadaly is constituted of a network of wires that, by electromagnetic induction, moves magnets located in the heart to simulate the wavering of the chest caused by breathing and to animate its frame. Hadaly's outer layer contains magnetized particles of iron that also conspire with the electric wires to make lifelike facial expressions and to convey to the touch the elasticity of human flesh. The gait (*démarche*) and balance of the human body is achieved through a system of communicating vessels filled with mercury. This system then regulates the dynamic interaction between induction and insulation and synchronizes movement by turning the electrical "fluid" on and off. An electrochemical battery powers the electromagnetic nervous system of the android to avoid cumbersome external wiring. In short, Hadaly, the new Eve, is a "dynamo-electric apparatus." And its "electro-magnetic motor" is primarily what allows it to simulate the human body and gait as never before.[70]

Hadaly would supersede Miss Alicia Clary by replacing her dull wits with the artificial intelligence supplied by two phonographs lodged in the chest, which would mix and play recordings from the unpublished works of the greatest nineteenth-century writers.[71] With potential immortality and untainted beauty, Hadaly's electromagnetic body is an idealization of life and cognition that provides Lord Ewald with a blank surface on which he can project his desires and, subsequently, the spark of vitality.

Skeptical but with nothing to lose, Lord Ewald decides to go ahead with the experiment. A few weeks later, Lord Ewald falls in love all over again with Miss Alicia Clary—without knowing that she is in fact the replica. The copy has surpassed the original and mended the heart of the aristocrat. Once he learns the truth, he first rebuffs the machine that has become the center of his affection, and then finally he gives in: "Phantom! Phantom! Hadaly! he cried, we must not part! Little credit to me for preferring your amazing miracles before that dull, deceptive, cold-hearted friend whom fate picked out for me! But let the heaven and earth take it as they will, I shall bury myself with you, my shadowy idol! I resign from the living [*Je donne ma démission de vivant*]—and let the age go about its business! For at last it's clear to me that set one beside the other, and it's the living girl who is the phantom!"[72] To love Hadaly is synonymous with giving up the living for the dead. Throughout *L'Ève future*, Hadaly has been constantly linked to the dead. Edison's automaton hides in an underground den referred to as the "tomb," wears a mourning veil, and travels in a "coffin." For Edison and Lord Ewald, Romantic idealism can only survive its bipolar disorder through the morbid embrace of a machine-produced illusion.

The Ghost in the Machine: Hysteria and Inductive Affinities

In *L'Ève future*, the modern and apparently down-to-earth engineer has taught the antiquated aristocrat a lesson about the artificial nature of love and desire. However, following Lord Ewald's concession, Edison, who has been alluding to a mystery since the beginning of the novel, finally reveals the secret of Hadaly's fabrication, the true catalyst of its seductive power. Edison has been able to perfect his simulation of the ideal woman with the aid of a magnetic somnambulist—Mrs. Annie Anderson, the widow of Mr. Edward Anderson, Edison's esteemed friend who had killed himself. As discussed in chapter 1, since the end of the eighteenth century, magnetic somnambulists had often been used to diagnose and cure illnesses. They were turned into a kind of healing machine, a condition that Friederike Hauffe, the famous seeress of Prevorst, rendered manifest when, in a magnetic dream, she described how to construct a "galvanic" machine that would restore her health and that she called her "*nerve-tuner*" (*Nervenstimmer*).[73]

After the demise of her husband, Mrs. Anderson contracted a severe neurosis that made her fall into states of continuous sleep. Edison took it upon himself to cure her with the help of animal magnetism. During the healing process, the great electrical engineer realized that he was also a powerful magnetizer who could telepathically communicate, even at a great distance, with his somnambulist. Edison implies that occult and technological media are basically the same when he mentions how at times she would reply to his thoughts using the telephone.[74]

Like the somnambulist in *Ursule Mirouët*, Mrs. Anderson starts to speak more eloquently and to manifest more acumen. In magnetic sleep, she seems to be inhabited by multiple other women. One in particular appears to have overtaken all the others, a superior intelligence named Sowana.[75] When Edison shares with Sowana his project to build Hadaly, she urges him to press on. She plans, "*occasionally*, TO INCORPORATE HERSELF WITHIN IT, AND ANIMATE IT WITH HER 'SUPERNATURAL' BEING."[76] Edison notices that Sowana, who has already taken possession of Mrs. Anderson's body, has expressed a conflicting emotion, a kind of sinister and vengeful joy at the prospect of "incorporating" the automaton that was to simulate Lord Ewald's ideal woman and thus prevent him from following the path that led his homologue, Mister Edward Anderson, to suicide. Sowana's "supernatural" incorporation provides the decisive clue to understanding Mrs. Anderson's illness, which recalls the haunting of Hauffe discussed above. Abraham, Torok, and Derrida would later diagnose this illness as a case of cryptic incorporation: following the violent death of her husband, Mrs. Anderson's mourning turned into severe melancholia, which provoked the incorporation of Sowana, an alter ego who is the idealized image of her husband's lover and who now wishes to take over the automaton's immortal body.[77] Mrs. Anderson's ghost originated in her husband's unconscious.[78] Sowana is the symptomatic manifestation of Mrs. Anderson's cryptic incorporation of the new Eve, the illusionary woman who had led Mr. Anderson, and many other nineteenth-century idealists, astray.

Edison comes close to this interpretation of Mrs. Anderson's haunting. For him, Sowana is the expression of the "Ideal."[79] She was able to transmit this quality to the automaton each time she possessed it and performed the fine-tuning necessary for Edison's creation to reach perfection. Through her alter ego's magical touch, Mrs. Anderson, a victim of the "Artificial" in desire, has prevented another suicide by turning a mere simulation of Miss Alicia Clary into the new Eve:

> The work is finished, and I can conclude that it has not resulted in an empty simulacrum [*vain simulacre*]. A soul has been added to it [. . .] to the voice, the gestures, the intonations, the smile, the very pallor of the living woman who was your love. *In her* all these qualities were dead, deceptive, degraded, because enslaved to vulgar, selfish reason; beneath their veils now lurks a feminine being who is, and perhaps

> always was, the true and rightful possessor of this extraordinary beauty, since she has shown herself worthy of it. In this way she who was the victim of the Artificial has at last redeemed the Artificial! She who was abandoned and betrayed by love turned degrading and obscene has grown into a vision capable of inspiring love at its most sublime! She who was blighted in her hopes, her health, and her prosperity by a wretched suicide has prevented another suicide.[80]

Edison takes his illusion for a reality. In fact, he has been duped: Sowana double-crossed him by letting him believe that the automaton had been transfigured into the "Ideal," when what she really did was use it to incorporate herself and seduce Lord Ewald. As hinted at by the similarity between his name and the first name of the late husband of Mrs. Anderson / Sowana, Ewald has become the new Edward. Mrs. Anderson's alter ego has won her back a husband substitute who shares the same bipolar disorder associated with idealism. However, this machination did not cure her melancholia. Soon after Lord Ewald leaves with his new girl toy to embark on a cross-Atlantic journey home, Mrs. Anderson passes away. The boat transporting her artificial body catches fire and claims (supposedly) her alter ego. The real Miss Alicia Clary also dies in the same accident. Unlike Edison, who considers his invention as a kind of modern-day *deus ex machina*, Lord Ewald did realize that he had been seduced by the ghost *in* the machine. Nevertheless, after the tragic accident, Ewald writes to Edison, "I grieve only for that shade,"[81] which implies that his bipolar disorder is incurable and that he will endlessly mourn the inevitable disappointment that comes with the pursuit of the "Ideal."

The magnetic and electromagnetic tropes are constantly invoked to reduce the supernatural aspects of Mrs. Anderson / Sowana's telepathic power and ability "to incorporate." As he tries to make sense of these strange phenomena, Lord Ewald gives at one point a late nineteenth-century press report on the electromagnetic wonders that are just about to overtake the world:

> It's already a remarkable thing that electric current can now transmit energy to great heights and over enormous, almost limitless distances. Indeed, if I'm to believe the reports that flood in from every direction, there is no doubt that tomorrow it will be used to spread through a thousand different networks the enormous blind energy, which always hitherto went to waste, of cataracts and torrents [. . .]. This trick is perfectly comprehensible, given the use of tangible conductors—magic highways—through which the powerful currents flow. But this SEMI-SUBSTANTIAL transmission of my living

> thought, how can I imagine it taking place, at a distance, *without conductor or wires, even the very thinnest?*[82]

Lord Ewald is describing how a dam works. The electric generator, or dynamo, transmits the mechanical force imparted by the flow of water to a turbine that in turn rotates a large magnet, which, through electromagnetic induction, generates electricity in a nearby conductor. This electricity can be transmitted potentially anywhere to power lights or, by reversing the process, to produce movement. Yet for Lord Ewald, the magic of the dynamo (or more precisely, of induction) does not measure up to the mystery of telepathic manifestations.

Edison's answer recalls Balzac's and Poe's fascination with clairvoyant states and other sympathies that appeared to bypass the conventional limits of space and time. With Edison, the magnetic somnambulist is momentarily substituted for her latest reincarnation, the hysteric (who became particularly famous during 1880s due to the highly publicized experiment of Jean-Martin Charcot at the Salpêtrière).[83] He writes,

> In the first place, replied the electrician, distance in these matters is nothing but a kind of illusion. Besides, you overlook here a number of facts recently verified by experimental science. For example, it is not just the nervous energy [*le fluide nerveux*] of a living being that can be transmitted over a distance, but the simple *virtue* of certain substances. Such substances can influence the human organism from afar without *ingestion, suggestion, or induction*. Here is an experiment that has been witnessed by a number of distinguished and very skeptical observers. A certain number of crystal jars are hermetically sealed and placed in envelopes; each contains a different drug, the same of which is concealed from me, the experimenter. I take one of them at random and hold it ten or twelve centimeters behind the head of . . . let's say, a hysteric. Within a few minutes the subject is seized with convulsions, vomits, sneezes, shouts aloud, or goes to sleep, according to the specific drug held behind his head at a distance.[84]

Then Edison speculates on the nature of these strange transfers that seem to happen without "*ingestion, suggestion, or induction*." His theory recalls the way Balzac at times apprehended similar phenomena in terms of a pervasive magnetic fluid and of a conception of the unity of natural forces, which resurfaces this time as the synthesis of the magnetic, electric, nervous, and hypnotic fluids into a "new fluid." Now that Edison has claimed the actual unity of these forces, he can explain how Sowana incorporated his electromagnetic machine. He can also conceive how, earlier, when Sowana touched Lord Ewald's hand, she was able to perceive Miss Alicia Clary through the channel of clairvoyance due to the network constituted by the common fluid of these three individuals and supported by the electromagnetic platform provided by the automaton:

> If some sort of *inductive affinity* can carry the vibrant influence of various drugs through the pores of the glass and the thicknesses of paper, to influence a patient in a state of hysterical supersensitivity, this is no more than a magnet does when its power passes through glass and through cloth to attract distant molecules or iron. And if *even vegetables and minerals* have an obscure sort of magnetism which without connectors [*inducteurs*] can cross distances and pass over obstacles to imprint their virtue on living beings, why should I be surprised if among three individuals of the same species, held together by a common electro-magnetic center, the various currents should so coincide as to produce telepathic communication [*les fluides, en un certain instant, soient devenus corrélatifs*]?[85]

The humorous repetition of the expression "without conductors [*inducteurs*]" in Lord Ewald's and Edison's question-and-answer session brings to mind a transfer without sender and receiver. Such undifferentiated communication of value is conveyed by Edison with the analogy of unyielding magnetic contagion. As the magnetic chain in Plato's *Ion* rendered manifest, such magnetic tropes are misleading, since magnetic contagion always needs a certain degree of discontinuity to propagate from one object to another. Furthermore, Edison ends up contradicting his monistic fluid theory when he speaks in the same passage of "*inductive affinity*" and when the unity of the fluid network linking Sowana, Lord Ewald, and Miss Alicia Clary is said to depend on a machine powered by electromagnetic induction.

In just a few paragraphs, Villiers displays an arsenal of magnetic analogies that both ignore and rely on the discovery of electromagnetism and the new metonymic model that came along with it. The opposition between noninduction and induction recalls the difference between the old and new conceptions of magnetic reproduction that Lodge and Rostand would later invoke in their magnetic analogies of life. Before Oersted, Ampère, and Faraday, magnetization would have represented life as the repetition of same. After them, through new images invoking electromagnetic induction, life is the product of a relation of contiguity between the two heterogeneous forces. As we will now see, Villiers's invocation of the metonymic model of "*inductive affinity*" is more significant in understanding how the long and detailed passages describing the construction of an electromagnetic animal in *L'Ève future* manifest a profound epistemological shift in the conceptualization of life that undermines the mechanical reduction of life and prefigures the invention of the unconscious.

Canguilhem on the Automaton and the Concept of Life

The machine has traditionally been the prevalent model for explaining the living. Georges Canguilhem locates the analogy of the living to a machine with the advent of the model of the automaton in the philosophical discourse. Before the automaton, the machine was a simple tool that could only represent the extension of animal or human muscular exertion, which rendered tautological the elucidation of the living with the machine. With the advent of the automaton, an impression of autonomy emerges that attenuates the continuity between machine and living being. At the same time, it produces a differentiated example to describe the function and functioning of the latter. A self-moving machine like a catapult possesses the capacity to store the energy necessary to propel its arm, creating a temporal gap between human and mechanical action substantial enough for Aristotle to identify the organs responsible for animal movement as parts, or *organa*, of the war machine.[86]

During the seventeenth century, the improvement and complexity of automata like the watch provoked a radicalization of the mechanical interpretation of the living, particularly via the Cartesian theory of the "animal machine." Conceived as a structure equivalent to the sum of its parts, the automaton continues to offer a concrete model suited to scientific analysis and to which knowledge of the living relates. Yet the machine is only a product of the living and, as Canguilhem argues, the finality of the living is much more uncertain than that of the machine. Unlike a machine, an organ is capable of *vicariance* (a healthy organ can take over the function of a defective one), *polyvalence* (organs can perform simultaneously more than one function), and self-regeneration.[87]

Over time, the autonomy of the automaton continued to increase due to the appearance of more sophisticated motors. Driven by the invention of the steam engine, the Industrial Revolution marked the intrusion into everyday life of powerful machines that looked more and more sovereign. The discovery of electromagnetic induction and the related inventions of the dynamo and electric motor started a second cycle of the Industrial Revolution, spearheaded by Edison's engineering breakthroughs and business ventures. During the 1870s, Villiers perceived that the instrumentalization of induction was making a critical contribution to the development of increasingly autonomous machines, and he began to write a novel organized around Edison and his creation of an electromagnetic automaton that could substitute for the living.

The formula responsible for the autonomy of the artificial yet improved version of "bourgeois Goddess" Miss Alicia Clary is the combination of an electromagnetic motor and its system of induction; the projections of a lovesick, bipolar idealist; and the incorporation of the automaton by an undead entity called Sowana. The various elements that must go into the construction of the electromagnetic animal allow Villiers to critique the prevalent mechanical thinking of the bourgeoisie that, at the height of the Industrial Revolution, reduced the living to a machine. Canguilhem criticizes this usurpation of mechanical thinking and the mechanization of thinking that comes along with it by arguing that the living precedes the machine. He inscribes the machine into the living. Similarly, Villiers uses his electromagnetic animal to demonstrate that the bourgeois model of the living is as mechanical as the model of a machine. He also suggests that an automaton able to fully emulate the living comes at the price of an occult intervention linked to death. For Villiers, the machine does not just come after the living; it is also situated in the space where life and death are metonymically linked. By drawing the detailed blueprint of a haunted electromagnetic automaton capable of substituting for the living, *L'Ève future* is an attempt to map this elusive space.

Psychoanalysis and the Electromagnetic Supplement

By bringing together electromagnetic interactions and technologies as well as hypnotic and telepathic phenomena linked to animal magnetism, hysteria, and mourning, Villiers's hybrid electromagnetic trope also provides an insightful representation of how cognition works. The new Eve incarnates the magnetic somnambulist and her docility as well as the keenness that had made an impact on Balzac's style. In Villiers, this Eve clearly represents the object of Romantic idealism that, after her husband's death, Mrs. Anderson incorporated and expressed through Sowana. The automaton also channels something occult, rendered by strange relations of contiguity at work in induction, mourning, and incorporation, that dupes the aristocrat, the magnetizer, and the engineer and that prefigures the psychoanalytical invention of the unconscious.

In 1895, nine years after the appearance of the final version of *L'Ève future*, Joseph Breuer and Sigmund Freud published *Studies on Hysteria*, their groundbreaking medical interpretation of hysteria, which paved the way for psychoanalysis and its new topology of the mind. In its theoretical chapter, Breuer imparts the electromagnetic apparatus powering the nervous system of Villiers's new Eve to the hysteric body—more particularly, to the revolutionary reconfiguration of the mind she had inspired:

> We ought not to think of a cerebral path of conduction as resembling a telephone wire which is only excited electrically at the moment at which it has to function (that is, in the present context, when it has to transmit a signal). We ought to liken it to a telephone line through which there is a constant flow of galvanic current and which can no longer be excited if that current ceases. Or better, let us imagine a widely-ramified electrical system for lighting and the transmission of motor power; what is expected of this system is that simple establishment of a contact shall be able to set any lamp or machine in operation. To make this possible, so that everything shall be ready to work, there must be a certain tension present throughout the entire network of lines of conduction, and the dynamo engine must expend a given quantity of energy for this purpose. In just the same way there is a certain amount of excitation present in the conductive paths of the brain when it is at rest but awake and prepared to work.[88]

When the intracerebral "tension" unexpectedly increases, or when the mental "dynamo" gets carried away due to certain affective circumstances associated with a bygone traumatic event, the hysteric cannot diffuse this sudden surge of electric energy via regular psychomotor pathways. The excess of intracerebral excitation can "short-circuit" and break through "at weak points in the insulation" of the nervous system to produce "electrical phenomena [. . .] at abnormal points." The breach provoked by an energy surge can influence an otherwise-insulated nervous system and can lead to "abnormal expression of emotion" or disrupt digestion.

Such new connections can endure and be revived or reelectrified under propitious affective conditions. This would explain how the hysteric suffers from "reminiscences" or how, after an initial and forgotten traumatic burst of electric energy, symptoms subsequently come back for no apparent reason. To discharge pathological excess tension, *Studies on Hysteria* prescribes a "talking cure" that relies heavily on hypnosis to bring back to consciousness what the patient had repressed at the moment of the trauma but that nevertheless lives on in her as a "foreign body" that triggers the repetition of her symptoms. The foreign body working within and marking the hysteric's body indicates the existence of "unconscious ideas" that operate outside of consciousness—that is, beyond the conventional channels of reflection, language, and memory.[89]

Hysterical repetition provided the clue that the mind was constituted by two contiguous systems (the conscious and the unconscious) that were fundamentally different yet interconnected. Goethe's logic of the compass had relied on magnetic bipolarity to convey such a metonymic link. At the end of the nineteenth century, Breuer identified and explored a similar relation of contiguity in the mind through an analogy now relying on an electromagnetic apparatus.

Akira Mizuta Lippit has argued that Breuer's ambivalent yet recurrent use of the "electromagnetic figure" shows that his theoretical account of hysteria and the unconscious also manifests a foreign body that short-circuits his attempt to downplay the role of analogy in his scientific discourse.[90] Since it functions through a mode of communication that ignores the self-conscious language used by Breuer's scientific discourse, the unconscious by definition resists clear-cut conceptualization and can only transpire through its figuration. Although Breuer claims that he does not want to identify the "splitting of the mind" with an electromagnetic system, he irresistibly comes back throughout his essay to this analogy to convey (as did Balzac, Poe, and Villiers before him) a dynamic system of invisible mental energies.

For Lippit, Breuer stumbled upon a new topology of the mind through the rapprochement of the mental and the electromagnetic systems that expressed what Jacques Derrida would later refer to as the logic of the supplement.[91] He paved the way for psychoanalysis with an electromagnetic trope that supplemented the unconscious; or, in other words, a trope that determined this foreign body even though it was invoked as a figurative stand-in for its elusive nature. The electromagnetic supplement establishes a metonymic relation with the unconscious that also informs Breuer's breakthrough concerning the mind. The unconscious supplements the conscious system. It establishes the subject while remaining radically different from conscious processes. This discovery, which was to have a tremendous impact on twentieth-century arts and sciences, emerged in its first instance through the analogy with an electromagnetic apparatus.

Since its emergence in the 1820s, electromagnetism progressively became an important model to conceptualize complex relations of contiguity and their logics of the compass and the supplement. To use a physics term, the advent of electromagnetism generated a ponderomotive force. As the Latin root of *ponder* suggests, this force had a deep impact on thought, empowering it to move beyond mechanical thinking as it redefined literary and scientific practices as well as notions such as life and cognition.

Conclusion

Bachelard's Electromagnetic Epistemology

The early history of magnetic and electromagnetic automata has shown how critical these real and imagined machines have been for giving form to and legitimizing the alternative modes of thought that I broadly called *electromagnetic thinking*. I have traced its development from its root in metonymic reasonings influenced by magnetic bipolarity to its nineteenth-century reformulation following the discovery of electromagnetic interaction. My focus on Faraday's induction apparatus and how it helped Balzac, Poe, and later Villiers explore various types of relations of contiguity has revealed the early and pervasive influence of this metonymic model during the nineteenth century.

I also argued that the same model played a crucial role in the reasoning that informed Einstein's elaboration of the theory of relativity. To bolster my claim that electromagnetic contiguity provides a unifying concept linking Balzac to Einstein and the broader metonymic shift that, as Schleifer pointed out, characterized cultural and scientific productions at the beginning of the twentieth century, I conclude this study with an epilogue that examines the critical impact that electromagnetic induction had on Bachelard's groundbreaking conceptions of history and the imagination as well as on Julien Gracq's remarkable essay discussing André Breton's surrealism.

Einstein's theory of relativity would profoundly influence the intellectual climate of the twentieth century by providing an exemplary case study for reevaluating the process of scientific discovery. Gaston Bachelard was one of the first epistemologists to draw extensively on Einstein's achievements to develop a new philosophy of science and history. For him, the theory of relativity not only transformed our conception of the universe but also signaled a "new scientific spirit,"[1] which originated in great part in Faraday's, Ampère's, and Maxwell's electromagnetic thinking. This new scientific spirit did not just operate within the accepted epistemological framework of the mid-nineteenth century. Instead of limiting itself to the material world of Newtonian mechanics, it extended the reach of science to "invisible and intangible *energy*."[2] Scientific investigation had previously derived mainly from empirical evidence and common sense. What electromagnetism revealed about the nature of the universe had very little to do with everyday experience. Einstein constructed a new and more accurate account of physical reality through bold reasoning and rigorous mathematical exploration that only later turned to empirical validation. For Bachelard, the rise of electromagnetism and the theory of relativity prove that scientific revolutions occur not through the continuous accumulation of knowledge but abruptly, through "epistemological breaks [ruptures]" triggered by unheralded ideas that stood fundamentally at odds with the accepted scientific framework of their time.[3]

The Bachelardian idea of an epistemological break was very influential in the development of "historical epistemologies" through the works of Louis Althusser and Michel Foucault, both of whom adapted it for their own purposes.[4] It also paved the way for Thomas Kuhn's paradigm shift theory. What is less known is the context of its first elaboration—namely, Bachelard's relativist and electromagnetic historical epistemology.

As it did in Poe's metonymic chains, in Balzac's description of inductive reasoning, and in Einstein's account of the discovery of the theory of relativity, electromagnetic induction plays a central role in Bachelard's historical epistemology.[5] According to Charles Alunni, Bachelard relied on this phenomenon and the formative role it played for Einstein's discoveries to define his own notion of cognitive induction.[6] In Bachelard's work, induction is a unifying concept describing a cognitive process of discovery common to scientific, philosophical, and literary practices that he calls *dynamic intuitions*: "The real world and the *dynamic determinism* that it implies calls for other *intuitions*, *dynamic intuitions* for which we should have a new philosophical vocabulary. If the word *induction* did not have already so many meanings, we would propose to apply it to these dynamic intuitions."[7] In his writings on epistemological breaks and the literary imagination, Bachelard did rely on the word *induction* to describe dynamic intuitions, and as we will see, its meaning was often informed by electromagnetism.

Through its canonization and subsequent reinterpretations, the expression *epistemological break* lost track of the electromagnetic context Bachelard invoked to conceptualize the radical discontinuities that marked the history of science. It hinged initially on a new interpretation of inductive reasoning informed by Einstein's induction apparatus. Bachelard writes: "There is no transition from the system of Newton to the system of Einstein. One does not proceed from the first to the second by amassing data, perfecting measurements, and making slight adjustments to first principles. What is needed is some totally new ingredient. It is a 'transcendental induction' and not an 'ampliative induction' that leads the way from classical to relativistic physics."[8] The term *ampliative* describes a type of induction that leads from the known to the unknown. As Bacon before them, Whewell and Mill associated their respective inductive methods with this process and thought that it was gradual and additive, each discovery paving the way for the next. Bachelard invokes the theory of relativity and the way it supplanted the Newtonian principles that had guided empirical research for the past two centuries to argue that the history of science is also marked by nonlinear breakthroughs.

Ampliative induction depends too much on a given empirical framework to trigger an "epistemological break." Einstein rejected the empirical constraint imposed by the physics of his time through a different type of inductive reasoning. As Einstein makes clear, and as Bachelard also knew,[9] a thought experiment that consisted of applying the principle of relativity to Faraday's induction apparatus is at the source of his world-changing discovery. This thought experiment resulted in what Bachelard calls a *transcendental induction*, which unveiled the more accurate principles of the theory of relativity.

Although he does not cite his source, which has led commentators to link it with Kant's transcendental deduction, the expression *transcendental induction* is a reference to another mid-nineteenth-century theory of inductive reasoning elaborated by the philosopher and theologian Joseph Gratry (1805–1872).[10] Like Whewell, Gratry supports an apriorism that guides empirical research as it uncovers the fixed unity of nature. Inductive reasoning uncovers this unity as it finds hidden relations among "contiguous" observations.[11] Gratry characterizes this metonymic imperative as "transcendent" because, as it connects finite observations to the infinite realm of God-given general principles, it crosses a fundamental threshold leading "from the same to the different."[12] Bachelard appropriated Gratry's transcendent view of induction to describe Einstein's discovery because, unlike the continuous and additive process invoked by ampliative induction, it signals a more dramatic break from its empirical point of departure.

The interaction of electricity, magnetism, and movement in Faraday's apparatus did not just induce the electric current that transformed the world at the turn of the twentieth century; it also *induced* a new scientific spirit. In physics, Einstein became one of the most outstanding examples of this new type of reasoning when he drew an analogy between electromagnetic induction and Galilean relativity to move the physics of his time beyond its own constraints. Bachelard extended the historical significance of this analogy to the philosophy of science when he formulated his nonlinear conception of scientific progress. The identification of a new type of cognitive induction, rendered manifest by the occurrence of "epistemological breaks" associated with electromagnetism, showed that epistemology itself was subject to profound change.

Bachelard's attraction to the multipurpose term *induction* to describe the "dynamic intuitions" of electromagnetic thinking resonates with Faraday's series of papers on his discovery of induction. In addition to defining the new electromagnetic meaning of *induction*, Faraday also uses the verb *to induce* in the sense of inductive reasoning.[13] In retrospect, Faraday's wide-ranging applications of *induction* begin to inform each other. This implicit link became more apparent in the Whewell-Mill debate when they argued over the proper interpretation of Faraday's method of discovery. Unlike Bachelard, neither Whewell nor Mill makes an explicit link between electromagnetic induction and inductive reasoning. It is another contemporary of Faraday, Balzac, who pioneered this link when, in his description of the volatile aspect of inductive reasoning, he substituted the traditional Newtonian image of the falling fruit for Faraday's induction apparatus.[14]

Psychic and Verbal Inductions

Bachelard also relies on electromagnetic induction to characterize a new spirit in modern literature. Dynamic intuitions stimulate both imaginative and rational works. Whereas in the latter it might lead through mathematical exploration to conceptual breakthroughs, in the former it inspires the formation of styles and images that affect the reader via "verbal induction": "Frequently, too, critics pay more attention to the word than to the sentence; they concentrate on the locution rather than the whole page. Their judgments are inevitably static, atomic. Few critics are prepared to test a new style by submitting to its *induction.* As I see it there ought to pass between writer and reader a kind of *verbal induction* sharing many of the characteristics of electromagnetic induction between two circuits. A book would then be a sort of psychic induction apparatus producing in the reader temptations to originality of expression"[15] (Bachelard's emphasis). In Bachelard, the term *induction* acquires the new implication of electromagnetic interaction to represent the elusive, dynamic intuitions at work in literary inventions. As in Poe's notion of "psychal fetters," induction allows Bachelard to think through a chain of transformations responsible for the transmission of a dynamic intuition from the author's imagination to the reader's. Like an electric current, the initial dynamic intuition produces a magnetic influence conducive to reverie and the elaboration of images. An effective style channels this potentially "moving" force and enables the process to reverse itself. The perceptive reader resembles a good conductor into which a magnetic style releases the electromotive force that had electrified the writer's imagination in the first place. As with the "transcendental induction" of an epistemological break, verbal induction marks the passage of a limit. It also establishes an affective continuity between writer and reader, which persists despite discursive mediation and participates in the tradition of the magnetic chain of enthusiastic *partage* inaugurated in *Ion*.

Bachelard implements his electromagnetic theory of communication in his own literary criticism. For instance, in *Lautréamont* (1939–63), he attempts "to relive the inductive force that runs throughout" *Les chants de Maldoror* (1869).[16] He attributes the electromotive force of *Les chants de Maldoror* to the singular violence of its style—more particularly, its animal imagery. Like the new scientific spirit, the new literary spirit moves beyond common language and conventional tropes to invent unsettling imagery conducive to psychic induction. There are also privileged images that owe their inductive force to the wealth of associations they naturally generate. For example, the word "root" (*racine*) is for Bachelard "an *inductive* word, a word that makes you dream, a word that comes to dream in us."[17] The idea of verbal induction allows Bachelard to think of a type of literary communication that goes beyond mimetic poetry or the mere duplication of sensory perceptions because it also transfers the actual movement responsible for the dynamic intuition that produced the image in the first place.[18]

After sensing the dynamic intuition contained in verbal induction, the task of modern literary criticism is to interpret and classify them according to the emotions and images they induce in the reader. In *Lautréamont*, Bachelard puts into practice the "metapoetics" he laid out at the end of *La psychanalyse du feu* (*The Psychoanalysis of Fire*, 1938), where he argued that literary analogies (which he broadly calls "metaphors") are not just isolated ornamentations but that they are organized in relation to one another.[19] The coordination of a group of metaphors gives way to a "syntax" that manifests the elusive forces structuring the imagination. Poets manipulate and sometimes break the rules regulating conventional metaphoric configurations, creating in turn a personal "syntax of metaphors" that differentiates them from each other. Bachelard develops an objective method for his literary criticism based on the identification of a poet's "syntax of metaphors," which functions for him like a "diagram which would indicate the meaning [*sens*] and the symmetry of his metaphorical coordinations, exactly as the diagram of a flower fixes the meaning and the symmetries of its floral action."[20]

Following *Lautréamont*, Bachelard's literary criticism actually turns away from mapping the syntax of metaphors of a single poet and concentrates instead on how metaphors from various authors form a coherent group by relating to a more fundamental image. He structures his studies according to four main groups that ancient philosophers, cosmologists, and alchemists considered the primordial elements: earth, water, air, and fire. Throughout the ages, the prominence of metaphors inspired by the four elements attests to their inductive power and to the intimate relation they share with the imagination. As fundamental images, the elements can both provide the raw materials for metaphoric formation and function as a "metaphor of metaphor" that expresses the secret movement and transformative "energies" of dynamic intuitions.[21] To gain access to a metaphor of metaphor, the literary critic (or "iconoclastic philosopher") has "to detach all the suffixes of beauty, to strive to find, behind the images that show themselves, the images that hide, to go to the very root of the imagining force."[22] Whether it describes the dynamic intuition of epistemological breaks or literary inventions, electromagnetic induction is itself a central "metaphor of metaphor" in Bachelard's oeuvre that helps him convey the dynamic and metonymic energies contained in language, reason, and the imagination.

Breton, Gracq, and the Magnetic Fields

Bachelard's metapoetics of the four elements inspired one of his younger contemporaries, Julien Gracq (1910–2007), to write the first in-depth study on the relation between electromagnetism, language, and the imagination. In his 1948 essay on André Breton—the charismatic father of surrealism who had jump-started one of the most influential art movements of the twentieth century with a work entitled *The Magnetic Fields* (1920)—Gracq demonstrates how electromagnetic induction is the fundamental image around which Breton's syntax of metaphors and aesthetics pivots.[23]

Gracq praises and builds upon Bachelard's metapoetics but disapproves of the manner in which it restricts the exploration of intimate cognitive dynamics and complexes to images solely associated with earth, water, air, and fire. He argues that the four elements are more representative of a "heavy" imagination hampered by their materiality than the "volatile" imagination of authors such as Rimbaud and Breton. The elusive and essentially dynamic domain associated with "shock," "transmutation," and the "contagious" at work in the volatile imagination better corresponds to imagery linked to electromagnetism.

The discovery of the fundamental and all-powerful nature of electricity and magnetism during the nineteenth century finally provided a model to legitimize cognitive phenomena that until then had lacked suitable representation.[24] Gracq appears unaware that electromagnetism deeply informs Bachelard's critical apparatus when he contends that the philosopher failed to add it to the four primordial elements as another key source of images particularly attuned to fathom the modern mind. Gracq goes on to demonstrate how electromagnetic induction supplies the key to Breton's avant-garde aesthetics and permeates the surrealist project.

His study begins by invoking the singular power of fascination Breton exerted on many writers and artists drawn to the surrealist cause. To these disciples, Breton functioned more as a charismatic or religious leader than as an "intellectual guide." For Gracq, electromagnetic induction provides the best analogy to explain the kind of "faith" or "blind following" inspired by Breton.[25] He also argues that he is not just using another scientific analogy to convey Breton's power of attraction. With the discovery of electromagnetic induction, certain primordial and elusive characteristics of the mind finally found the right image to reveal themselves: "Such an image seems endowed with a revealing power on this part of ourselves where involuntary attractions and repulsions take root. It seems that we are in the presence of a pure motive image, quite ancient, born from a state of need, but relentlessly seeking since its birth a concrete material counterpart, which it feels in advance the right to ask from the outside world and be certain to discover, despite its disappointments, either first of all in Mesmer's 'animal magnetism' or in the field of chemistry, with Goethe's 'elective affinities.'"[26] Electromagnetic induction supplies this "concrete material" (or fundamental image) to represent aspects of the mind that Mesmer's "animal magnetism" and Goethe's "elective affinities" imperfectly express. As "a pure motive image" (or metaphor of metaphor), it conveys secret movements of the mind. The rise of animal electromagnetism traced in chapters 1 and 2 was a branch of mesmerism that began to supply such a fundamental image to Balzac and Poe.

As Valéry did in his dynamo thought experiment, Gracq contrasts the advent of electromagnetic induction during the nineteenth century with the "purely mechanical psychology" that preceded it. He compares the idealized "mechanics of the passions" as depicted in the neoclassical plays of Jean Racine (1639–1699) to the "induced phenomena" shaping the characters' interpersonal relations in the realist and psychological novels of Fyodor Dostoevsky (1821–1881): "Seventeenth-century 'man' can be understood by us, [. . .] but the *current* never passes through him to let all the procession of induced phenomena appear—nothing can be glimpsed from this sacred dance that improvises itself through the contact between two beings and whose subsequent behavior will continue to develop until the exhaustion of variations—never through this deified monad, the *whole* of man, which is to be a self for ever caught in the magnetic whirlpool of a system of 'self' [. . .] can take on the least amount of reality"[27] (Gracq's emphasis). Racine's *anatomical* psychology reduces the complexity of interpersonal relationships exhibited by Dostoevsky's *field* psychology. The interaction of selves reveals a magnetic field in which they participate, as that field carries their invisible influences. As with Breton and the surrealist group, and as first described in Plato's *Ion* by the magnetic chain, the field of influence of charismatic leaders becomes particularly manifest in the unrestrained enthusiasm of their followers.

Like the bodily process described in Balzac's electromagnetic analogy of mesmerism and Faraday's induction apparatus, the underlying factors in group psychology are invisible and can only be studied indirectly:

> Compared to this purely mechanical psychology, Dostoevsky unquestionably belongs to the other pole: as soon as we open a book like *The Possessed* [. . .] we find ourselves thrown amidst an irradiated world, crossed, thanks to the agency of a self who is an exceptionally good conductor, by influxes and lines of force. The central figure of Stavrogin is not endowed with a particularly differentiated mental structure, but with both the impenetrable and fascinating characteristics of electric manifestations: probably for the first time with such clarity [. . .] are we presented with a being whose reality, though larger than life, expressly and solely reduces itself to the *trail* [*sillage*]

it leaves, to phenomena of turbulence, accelerated decomposition and recomposition that it generates on its path.[28]

Gracq describes the emergence of inductive literary representation in Dostoevsky's novel by invoking the indirect and metonymic method of recording implemented in a cloud chamber (fig. 11). The cloud chamber was one of the first instruments built to visualize and identify subatomic particles. Maxwell's electromagnetic wave theory paved the way for the discovery of radiation such as radio waves and X-rays as well as the discovery of radioactive materials such as uranium. Radiations and radioactivity in turn yielded an unprecedented look into the hidden world of the atom. Once J. J. Thomson discovered the electron in 1897, the identification of other electrically charged subatomic particles soon followed, signaling the dawn of nuclear and quantum physics.[29] The scale and nature of subatomic particles make it impossible to detect them directly. Physicists developed experiments to identify these particles by the metonymic traces, or "tracks," that their passages left in another medium such as gas. In 1911 Charles T. R. Wilson (1869–1959) invented the cloud chamber, where the interaction between the electric charge of traveling subatomic particles and water molecules results in vapor trails visible to the eye. Due to their various electric charges, subatomic particles react differently under the influence of a magnetic field applied to the cloud chamber, and they consequently draw unique individual paths, which scientists use to identify them.

For Gracq, the impassioned members of the surrealist group and Dostoevsky's psychological novel render manifest a kind of cloud chamber. Breton and Stavrogin are electrically charged particles whose secret energy profoundly marked their respective environments. Under the influence of a medium akin to a magnetic field, their invisible interpersonal relations leave characteristic vapor trails that supply metonymic clues with which to identify the nature of their psychic inductions.

The rise of modern psychology has been closely tied to discoveries in electromagnetic science. Mesmer's protohypnotic therapy, animal magnetism marked the beginning of "talking cures." Nineteenth-century literary and scientific exploration of nonanatomical psychology culminated in 1895 with the invention of psychoanalysis and Gustave Le Bon's groundbreaking work on group psychology. Psychoanalysis, which practices the indirect and often metonymic method of inductive reasoning known as *free association*, was initially informed by an electromagnetic apparatus. In *Psychologie des foules* (1895), Le

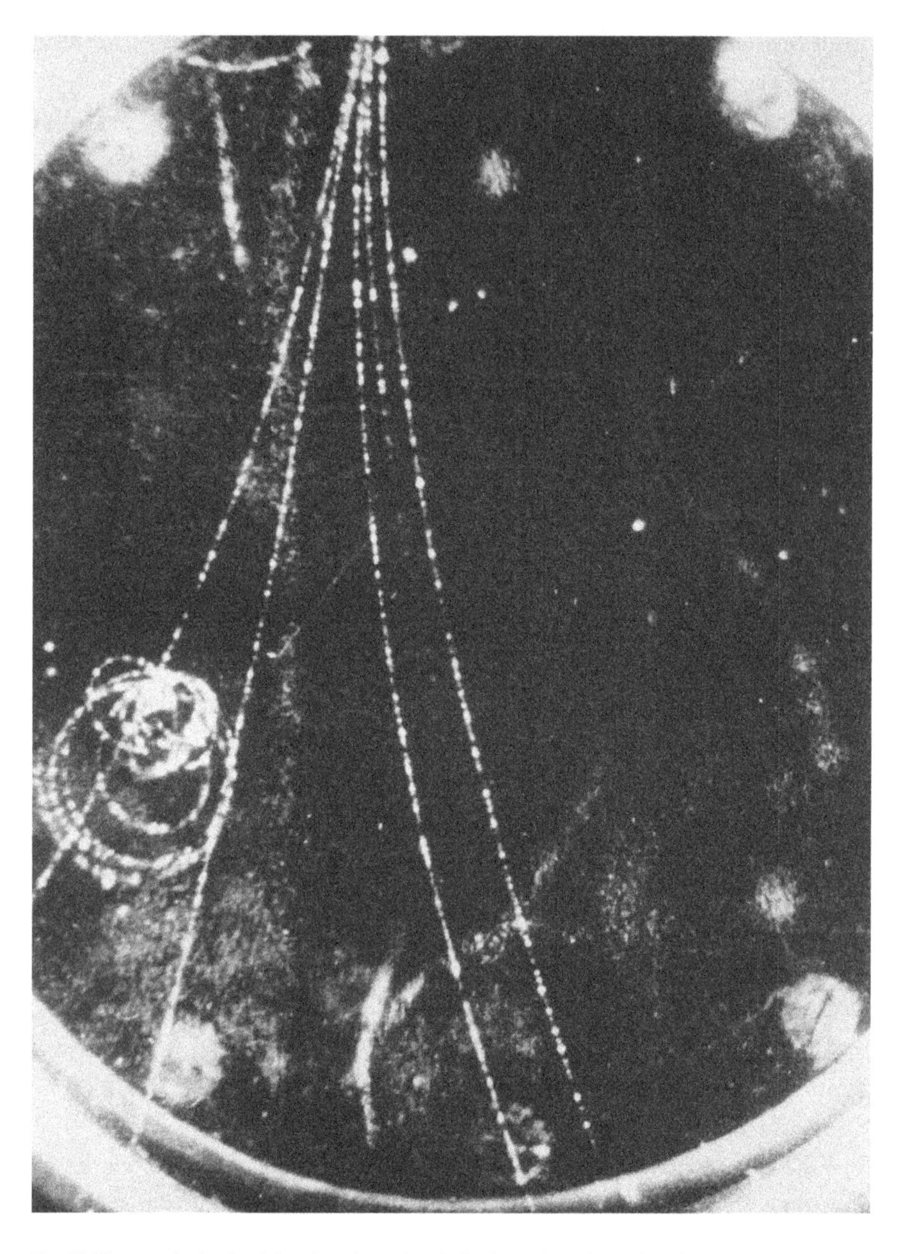

Fig. 11 Photograph of a cloud chamber where electrically charged atomic particles interact with a magnetic field and become visible through their distinctive vapor trails. This photograph shows a "shower" of three electrons and three positrons produced by a cosmic ray. The electrons bend to the left in the chamber's magnetic field, the positrons to the right. This photo was taken by Carl Anderson on top of Pike's Peak, Colorado, and published in 1936. Credit: Carl Anderson / Science Source.

Bon studies the suggestibility of groups and phenomena of "mental contagion" in terms of hypnosis and magnetism.[30] Le Bon's career path might have inspired Gracq's striking image of the cloud chamber associating group psychology with nuclear physics. The same year that he published his study on crowds, Le Bon became interested in radiation and the emerging field of atomic physics. In 1905 he published a best seller popularizing the latest findings about radiation and the enormous energy stored in the atom.[31]

For Gracq, Breton's "magnetic personality" radiates beyond the fervor of his disciples and creates its most distinct vapor trails in his writing.[32] In *Psychologie des foules*, Le Bon's discussion of the major impact that certain images, words, and formulae have on crowds can also help explain the transition in Gracq's thinking from Breton's power of fascination to surrealist verbal induction. For Le Bon, the lack of critical distance characteristic of crowds makes them particularly susceptible to images. Words and formulae possess the ability to summon such images, and they usually do so regardless of meaning. Notoriously vague words such as *democracy*, *socialism*, *equality*, and *freedom* can have different meanings depending on the place and historical context of their utterance—yet they always impress crowds.[33]

Meaning is not the main component that stirs the imagination. There is a "truly magical power" that allows certain terms to evoke images connected to "various unconscious aspirations." To convey the magical function of words and formulae, Le Bon relies on electromagnetic induction as implemented in early telephony: "A word is only the electric call button [*bouton d'appel*] that makes images appear."[34] At the end of the nineteenth century, a phone call started by pressing a call button that closed an electrical circuit. Via the activation of an electromagnet, this sent a signal to the switchboard operator.[35] As another application of electromagnetic interaction, the electric call button prefigures what Bachelard would later call *verbal induction*. The similar ideas of an electric call button and verbal induction allow Le Bon and Bachelard to refer to a mode of communication operating beyond the channel responsible for the transmission of meaning, and considered mainly as metaphysical, until the related discoveries of electromagnetic and unconscious phenomena. Whereas in Bachelard this elusive mode of communication can stimulate "original expression" and the process of discovery, for Le Bon it manipulates crowds.

Bachelard's appreciation of verbal induction comes partly from his fondness for surrealist authors, who, as Gracq demonstrates in the case of Breton, had dedicated their writings to exploring and harnessing the elusive mode of communication epitomized by electromagnetic interaction. From the title of his first properly surrealist work, *The Magnetic Fields*, to his constant use of electromagnetic imagery, Breton leaves no doubt as to what informs his "volatile" imagination. Gracq spends several pages systematically citing and commenting on the numerous instances where Breton relies on electromagnetism to convey his favorite subject of psychic and verbal inductions. Among the related phenomena structuring Breton's syntax of metaphors are corona discharge (*aigrette*), lighting, sparks, wireless telecommunication, induction, electric conductors, the electroscope, invisible rays, lines of force (*gerbe de rayons*), the field, geomagnetism, polarity, the compass, short circuiting, magnetization, and induced electricity.

As a poet and theorist, Breton structures surrealism around the guiding image of electromagnetism to legitimize a domain of investigation that eludes traditional mechanical thinking. This was first rendered manifest in *The Magnetic Fields* and its exploration of "automatic writing." Gracq writes: "By loosening up as much as possible the mechanical rules joining words, by freeing them from the banal attractions of logic and habit, by letting them 'fall' in an interior vacuum [. . .], he will observe and blindly follow the secret magnetic attractions between them, he will let 'the words make love' and finally let a strange *world* be recomposed through their freedom. [. . .] he will assist from outside, as a spectator, at the spontaneous elaboration of this continuous magic, confining himself to signing by a frivolous act that will always include a dose of humor, the most accomplished crystallizations."[36] Automatic writing seeks to emulate a state similar to what Le Bon perceives in crowds, where the affective power of words and formulae supplants their meaning. Through long streams of uninterrupted writing, the surrealist scribe becomes the heir of the somnambulist-writer by minimizing awareness and turning into the spectator of unconscious aspirations. Involuntary and unexpected word association that resists common logic and the habitual production of meaning can happen during automatic writing. The writer then sorts out the fortuitous couplings, keeping only those that communicate the most striking images. This indirect surrealist method actively searches for new "electric call buttons" to explore an affective field that operates beyond the channel of meaning.

For Gracq, even though verbal induction does not communicate well-defined meaning, it can transmit knowledge. Like Bachelard's dynamic intuition, this knowledge originates in the nascent state of thought and in the primordial movement and "vibration" that gives birth to thought:

> The surfacing of a thought into consciousness always participates more or less in the moving characteristic of birth—intellectual, indeed, in its essence, it is indefinably mixed with a colored fringe of affectivity, eminently vibrational, able to send out waves to neighboring zones, it is a motive force and magnetic potential capable of infinite and abrupt variations. Well, it is clear that this nascent state of a thought, vague, proteiform, but endowed with a considerable affective charge (thought turned into vibration), is the one among all that appears to be the most capable of communication. Certain gazes or gestures from an inspired actor, certain *movements* from a great poet, before all elucidation of its content, arouse our sensibility—unlike that mental pittance, so easily prey to counterfeiters, that stirs "the circulation of ideas"—to *contagious* states of thought.[37] (Gracq's emphasis)

There is an affective dimension to thought that should not be mistaken for the "mental pittance" holding sway over crowds. Thought does not consist solely of coding, decoding, and managing content; it also involves the communication of its constitutive dynamic. The transmission of this vague and "proteiform" knowledge through "contagious states of thought" has mainly been the domain of artistic production. Like Bachelard, Gracq describes the communication of this dynamic knowledge as an electromagnetic induction between two circuits. He also connects it to a theory of knowledge whose name Paul Claudel coined as *co-naissance*:

> Attached to this moment of mystery where the spirit becomes pure by its blossoming, we know in the end that poetry asks for nothing more, by the breaking off of all habitual associations by means of the image, to provoke artificially this nascent state by trying to make us see each object in the light of world creation, and as though for the first time. In this state of poetic grace, whether an illusion or not, *the current seems to pass from conscience to conscience without obstacle*, a spontaneous state of resonance happens which, beyond the impression of communication, produces the impression of "knowledge" [*co-naissance*] to use Claudel's expression.[38] (my emphasis)

Gracq does not comment any further on Claudel's notion of co-naissance. *Co-naissance* is a pun on the French word for knowledge, *connaissance*, and means "cobirth," or "born with." For Claudel, the act of knowing gives birth to both the object and the subject. To know an object implies informing and being informed by it. The detached subject is then a myth.[39] Born out of a mutual and concurrent induction, the subject and the object share a metonymic link that is the necessary precondition for the formation of knowledge and, according to Gracq, electromagnetic thinking.

Magnetism and electromagnetism shaped and reshaped the history not only of literature but also of scientific thinking. When Valéry claimed that the dynamo would baffle such great minds as Archimedes, Galileo, Descartes, or Newton, this underscored the epistemological break that Bachelard attributed to the emergence of electromagnetic thinking. Electromagnetism unveiled a relation of contiguity that caused profound and connected changes in literary and scientific practices because it offered a fresh way to explore puzzling relations between object and subject, mind and matter, and life and death. From Plato's *partage* to Goethe's logic of the compass, magnetic transformation and bipolarity paved the way for these changes. During the nineteenth century, electromagnetic induction quickly became one of the most important motors for conceptual as well as technological invention. In addition to powering new machines such as the telegraph, dynamo, and power grid, induction also empowered alternative ways to investigate the world. From the literary experimentations of Poe, Balzac, Villiers, and Breton to the creative leaps of Faraday and Einstein, electromagnetic induction legitimized imaginative modes of reasoning based on a more acute sense of interconnection and a renewed interest in how metonymic relations could reveal the order of things.

Notes

Introduction

1. Paul Valéry, "Le bilan de l'intelligence," in *Œuvres*, Bibliothèque de la pléiade, vol. 1 (Paris: Gallimard, 1957), 1060–61.

2. Valéry's characterization of Newton is typical of the eighteenth and nineteenth centuries. But scholars have recently demonstrated that it is reductive to just call him a mechanical thinker. See chapters 4 and 5 in Robert Markley, *Fallen Languages: Crises of Representation in Newtonian England, 1660–1740* (Ithaca: Cornell University Press, 1993).

3. Sam Halliday, *Science and Technology in the Age of Hawthorne, Melville, Twain, and James: Thinking and Writing Electricity* (New York: Palgrave Macmillan, 2007); Paul Gilmore, *Aesthetic Materialism: Electricity and American Romanticism* (Stanford: Stanford University Press, 2009); Jennifer L. Lieberman, *Power Lines: Electricity in American Life and Letters, 1882–1952* (Cambridge: MIT Press, 2017).

4. This shift of focus from electricity to electromagnetism has already proven fruitful in the field of art history. Linda Dalrymple Henderson and Douglas Kahn have traced how various early interpretations of electromagnetism have significantly shaped early twentieth-century visual and sound art. My study extends their electromagnetic genealogy of modern art to literature and back to the first half of the nineteenth century. Linda Dalrymple Henderson, *Duchamp in Context: Science and Technology in the Large Glass and Related Works* (Princeton: Princeton University Press, 1998); Douglas Kahn, *Earth Sound Earth Signal: Energies and Earth Magnitude in the Arts* (Berkeley: University of California Press, 2013).

5. Devin Griffiths, *The Age of Analogy: Science and Literature Between the Darwins* (Baltimore: Johns Hopkins University Press, 2016), 7.

6. Michael Friedman, "Kant—Naturphilosophie—Electromagnetism," in *Hans Christian Ørsted and the Romantic Legacy in Science Ideas, Disciplines, Practices*, ed. Robert Michael Brain, R. S. Cohen, and Ole Knudsen (Dordrecht: Springer, 2007). For the specific influence of Schelling's notion of polarity on Oersted via the experiments of Ritter, see Roberto de Andrade Martins, "Ørsted, Ritter, and Magnetochemistry," in Bain, Cohen, and Knudsen, *Hans Christian Ørsted*, 339–85. See also Kenneth L. Caneva, "Physics and Naturphilosophie: A Reconnaissance," *History of Science* 35, no. 107 (1997); Friedrich Steinle, "Romantic Experiment? The Case of Electricity" (paper presented at the Ciencia y romanticismo, Maspalomas, 2002).

7. Ronald Schleifer, *Modernism and Time: The Logic of Abundance in Literature, Science, and Culture, 1880–1930* (Cambridge: Cambridge University Press, 2000), 13–15, 185.

8. Roman Jakobson, "Two Aspects of Language and Two Types of Aphasic Disturbances," in *Fundamentals of Language* (The Hague: Mouton, 1956); George Lakoff and Mark Johnson, *Metaphors We Live By* (Chicago: University of Chicago Press, 1980); Gerard Steen, "Metonymy Goes Cognitive-Linguistic," *Style* 39, no. 1 (2005); *Metonymy*

in Language and Thought, ed. Klaus-Uwe Panther and Günter Radden (Amsterdam: Benjamins, 1999). *Metaphor and Metonymy at the Crossroads: A Cognitive Perspective*, ed. Antonio Barcelona (Berlin: De Gruyter, 2012); Boguslaw Bierwiaczonek, *Metonymy in Language, Thought and Brain* (Sheffield: Equinox, 2013); Jeannette Littlemore, *Metonymy: Hidden Shortcuts in Language, Thought and Communication* (Cambridge: Cambridge University Press, 2015); *Conceptual Metonymy: Methodological, Theoretical, and Descriptive Issues*, ed. Olga Blanco-Carrión, Antonio Barcelona, and Rossella Pannain (Amsterdam: John Benjamins, 2018). See also Klaus-Uwe Panther and Linda Thornburg, "Metonymy," in *The Oxford Handbook of Cognitive Linguistics*, ed. Dirk Geeraerts and H. Cuyckens (Oxford: Oxford University Press, 2007).

9. In the wake of Einstein's extension of field theory to gravity, Roman Jakobson describes the two poles as "gravitational poles." The French translation of the essay refers to them as "cardinal poles," which brings to mind the magnetic compass. Jakobson, "Two Aspects of Language," 76–78; Roman Jakobson, *Essais de linguistique générale* (Paris: Éditions de Minuit, 1963), 62.

10. Sebastian Matzner, *Rethinking Metonymy: Literary Theory and Poetic Practice from Pindar to Jakobson* (Oxford: Oxford University Press, 2016), 165. See also Michael Silk, "Metaphor and Metonymy: Aristotle, Jakobson, Ricoeur, and Others," in *Metaphor, Allegory, and the Classical Tradition: Ancient Thought and Modern Revisions*, ed. G. R. Boys-Stones (Oxford: Oxford University Press, 2003).

11. Matzner, *Rethinking Metonymy*, 25–53, 154–65.

12. See, for instance, Kenneth Burke, "Four Master Tropes," *Kenyon Review* 3, no. 4 (1941). Gérard Genette, "La rhétorique restreinte," *Communications*, no. 16 (1970): 163.

13. Charles Lock, "Debts and Displacements: On Metaphor and Metonymy," *Acta Linguistica Hafniensia* 29, no. 1 (1997): 323–24.

14. David Hume, "An Abstract of [. . .] a Treatise of Human Nature," in *A Treatise of Human Nature*, ed. David Fate Norton and Mary J. Norton (Oxford: Oxford University Press, 2000), 4; David Hume, *A Treatise of Human Nature*, ed. David Fate Norton and Mary J. Norton (Oxford: Oxford University Press, 2000), 416. For an insightful study on the profound impact of associationism on the history of literature, see Cairns Craig, *Associationism and the Literary Imagination: From the Phantasmal Chaos* (Edinburgh: Edinburgh University Press, 2007).

15. Hume, *Human Nature*, 14.

16. See query 31 in Isaac Newton, *Opticks: or, A Treatise of the Reflections, Refractions, Inflections and Colours of Light*, 4th ed. (London: William Innys, 1730). See also the scholium following proposition 69 in Isaac Newton, *The Mathematical Principles of Natural Philosophy*, trans. Andrew Motte, vol. 1 (London: Benjamin Motte, 1729). A few years after Hume, another influential associationist philosopher, David Hartley, invoked gravity, electricity, and magnetism in his discussion on "Attraction." David Hartley, *Observations on Man, His Frame, His Duty, and His Expectations*, vol. 1 (London: S. Richardson, 1749), 28.

17. Lines 533d–533e. Translation (with minor modifications) taken from Plato, *Ion*, accessed June 2019, http://classics.mit.edu/Plato/ion.html.

18. Jean-Luc Nancy, "Le Partage des Voix," in *Ion*, ed. Jean-François Pradeau and Édouard Mehl (Paris: Ellipses, 2001).

19. Cited in Koen Vermeir, "Magnetic Theology as a Baroque Phenomenon" (Baroque Science Workshop, University of Sydney, 2008).

20. Ibid., 12.

21. Similar depictions of the magnetic chain appear in other works from the same era. In 1637 Samuel Ward includes it in the frontispiece of his book on "magnetic theology." See also chapter 5 in John Peacock, *The Look of Van Dyck: The Self-Portrait with a Sunflower and the Vision of the Painter* (Aldershot, U.K.: Ashgate, 2006).

22. David C. Gooding, "From Phenomenology to Field Theory: Faraday's Visual Reasoning," *Perspectives on Science* 14, no. 1 (2006).

23. Thomas K. Simpson, *Figures of Thought: A Literary Appreciation of Maxwell's Treatise on Electricity and Magnetism* (Santa Fe: Green Lion, 2005).

24. Michael Faraday, "On Lines of Magnetic Force [. . .]," in *Experimental Researches in Electricity* (London: R. and J. E. Taylor, 1839–55), 349.

25. Faraday, "On Some Points of Magnetic Philosophy," in *Experimental Researches in Electricity*, 571–72, 574.

26. For a discussion of this diagram, see Gooding, "Visual Reasoning," 58.

27. James Clerk Maxwell, *A Treatise on Electricity and Magnetism*, vol. 1 (Oxford: Clarendon Press, 1873), ix.

28. Oliver Heaviside (1850–1925) is mainly responsible for extracting and arranging these equations from Maxwell's work. Bruce J. Hunt, *Pursuing Power and Light: Technology and Physics from James Watt to Albert Einstein* (Baltimore: Johns Hopkins University Press, 2010), 114.

29. See James Clerk Maxwell, "On Action at a Distance," in *The Scientific Papers of James Clerk Maxwell* (Cambridge: Cambridge University Press, 1890), 321–22. And chapter 20 in James Clerk Maxwell, *A Treatise on Electricity and Magnetism*, vol. 2 (Oxford: Clarendon Press, 1873).

30. Maxwell, *Electricity and Magnetism*, vol. 2, 390–91.

31. According to Gooday, the development and commercialization of magnet-free rechargeable batteries around the same time also contributed to downplaying the importance of magnetism. Graeme Gooday, *Domesticating Electricity: Technology, Uncertainty and Gender, 1880–1914* (London: Pickering & Chatto, 2008), 48–50.

32. Franklin's kite experiment proved that lightning is an electric phenomenon and that its awesome power could be tamed. The Leyden jar provided a device that concentrated and stored electrostatic energy to an unprecedented level, which made electricity easier to study and was the subject of sensational experiments involving sparks and shocks. Volta demonstrated that chemical interaction could generate a steady "current," unveiling a whole new domain of electric investigations and applications. For a recent insightful study on the subject, see James Delbourgo, *A Most Amazing Scene of Wonders: Electricity and Enlightenment in Early America* (Cambridge: Harvard University Press, 2006).

33. Patricia Fara, *Sympathetic Attractions: Magnetic Practices, Beliefs, and Symbolism in Eighteenth-Century England* (Princeton: Princeton University Press, 1996); Pierre Juhel, *Histoire de la boussole: L'aventure de l'aiguille aimantée* (Versailles: Editions Quae, 2013); Stephen Pumfrey, *Latitude and the Magnetic Earth* (New York: MJF Books / Fine Communications, 2006).

34. Aphorism 129 in Francis Bacon, *The New Organon*, ed. Lisa Jardine and Michael Silverthorne (Cambridge: Cambridge University Press, 2000), 100.

35. Michel Serres, *Hermès IV: La distribution* (Paris: Éditions de Minuit, 1977). And chapter 2 and the beginning of part 2 in Michel Serres, *Feux et signaux de brume, Zola* (Paris: Grasset, 1975).

36. Sadi Carnot, *Réflexions sur la puissance motrice du feu et sur les machines propres à développer cette puissance* (Paris: Bachelier, 1824).

37. Bruce Clarke, *Energy Forms: Allegory and Science in the Era of Classical Thermodynamics* (Ann Arbor: University of Michigan Press, 2001); Barri J. Gold, *ThermoPoetics: Energy in Victorian Literature and Science* (Cambridge: MIT Press, 2010); Sydney Lévy, "Balzac et la machine-*peau de chagrin*," *Théorie, littérature, enseignement* 21 (2003).

38. "The second era of rapid industrialization is not as well understood or defined as the first, but when future historians probe its phenomena and structure, electric power—especially high-voltage transmission and the universal system of electric light and power made known to the world at Frankfurt and Niagara Falls—will figure intricately and prominently in their interpretations and representations." Thomas Parke Hughes,

Networks of Power: Electrification in Western Society, 1880–1930 (Baltimore: Johns Hopkins University Press, 1983), 175.

39. John Tresch, *The Romantic Machine: Utopian Science and Technology After Napoleon* (Chicago: University of Chicago Press, 2012), 103–4, 37.

40. Maxwell, *Electricity and Magnetism*, 2, 162.

Chapter 1

1. This tale has received little scholarly attention aside from a handful of articles that mention it in relation to optics. Rae Beth Gordon, "Poe: Optics, Hysteria, and Aesthetic Theory," *Cercles* 1, no. 49 (2000): 49–60; Susan Elizabeth Sweeney, "The Magnifying Glass: Spectacular Distance in Poe's 'Man of the Crowd,'" *Poe Studies* 36, no. 1 (2003): 3–17; William J. Scheick, "An Intrinsic Luminosity: Poe's Use of Platonic and Newtonian Optics," *Southern Literary Journal* 24, no. 2 (1992): 90–105. Critics tend to gloss over the tale's initial success, the unusual effort Poe invested to publish it in England, and his plan to include it in the next volume of his collected tales. Burton R. Pollin, "'The Spectacles' of Poe—Sources and Significance," *American Literature* 37, no. 2 (1965): 185–90.

2. Edgar Allan Poe, "The Spectacles," in *Collected Works of Edgar Allan Poe: Tales and Sketches, 1843–1849*, ed. Thomas Ollive Mabbott (Cambridge: Belknap Press of Harvard University Press, 1978), 883–918, https://www.eapoe.org/works/mabbott/tom3t007.htm. 6/2019.

3. *Oxford English Dictionary* dates one of the first appearances of the word *psychal* to Poe's usage and defines it as "relating to the psyche; spiritual, psychical, psychological."

4. Edgar Allan Poe, "Marginalia: Installment V: Graham's Magazine, March 1846," in *The Collected Writings of Edgar Allan Poe*, ed. Burton R. Pollin (New York: Gordian Press, 1985), 253–62, http://www.eapoe.org/works/misc/mar0346.htm. 6/2019.

5. Suzanne Stern-Gillet, "On (Mis)interpreting Plato's *Ion*," *Phronesis* 49, no. 2 (2004): 169–201.

6. Percy Bysshe Shelley, *Essays, Letters from Abroad, Translations and Fragments*, ed. Mary Wollstonecraft Shelley, vol. 1 (London: Edward Moxon, 1840), 5–6, 16, 19, 44–47.

7. Arthur O. Lovejoy, *The Great Chain of Being: A Study of the History of an Idea* (Cambridge: Harvard University Press, 1936). Whereas the image of the chain begins with Plato, the notions of continuity and gradation that link everything in nature become more concrete in Aristotle: "Nature proceeds little by little from things lifeless to animal life in such a way that it is impossible to determine the exact demarcation, nor on which side thereof an intermediate form should lie." Cited in William F. Bynum, "The Great Chain of Being After Forty Years: An Appraisal," *History of Science* 13 (1975): 4. See also Lovejoy, *History of an Idea*. Renaissance humanist philosopher Marsilius Ficinus wrote an influential neo-Platonic interpretation of *Ion* linking Plato's magnetic chain with Aristotle's gradations. Plato, *Ion*, ed. Jean-François Pradeau and Édouard Mehl (Paris: Ellipses, 2001).

8. Nick Hopwood, Simon Schaffer, and Jim Secord, "Seriality and Scientific Objects in the Nineteenth Century," *History of Science* 48, no. 3–4 (2010): 252.

9. Ibid., 266–67. And in the same volume, Chitra Ramalingam, "Natural History in the Dark: Seriality and the Electric Discharge in Victorian Physics," 371–398.

10. Adam Frank, "Valdemar's Tongue, Poe's Telegraphy," *ELH* 72, no. 3 (Fall 2005). See also Christina Zwarg, "Vigorous Currents, Painful Archives: The Production of Affect and History in Poe's 'Tale of the Ragged Mountains,'" *Poe Studies* 43, no. 1 (2010).

11. For instance, Richard Menke, *Telegraphic Realism: Victorian Fiction and Other Information Systems* (Stanford: Stanford University Press, 2008); Paul Gilmore, *Aesthetic Materialism: Electricity and American Romanticism* (Stanford: Stanford University Press, 2009); Aaron Worth, *Imperial Media: Colonial Networks and Information*

Technologies in the British Literary Imagination, 1857–1918 (Columbus: Ohio State University Press, 2014).

12. Bruce Mills, *Poe, Fuller, and the Mesmeric Arts: Transition States in the American Renaissance* (Columbia: University of Missouri Press, 2006). Bruce Mills, "Mesmerism," in *Edgar Allan Poe in Context*, ed. Kevin J. Hayes (Cambridge: Cambridge University Press, 2013); Zwarg, "Vigorous Currents.'"

13. Alan Gauld, *A History of Hypnotism* (Cambridge: Cambridge University Press, 1992).

14. Robert Darnton, *Mesmerism and the End of the Enlightenment in France* (Cambridge: Harvard University Press, 1968), viii.

15. The next two paragraphs are indebted to the historical account in Wilfried Andrä and Hannes Nowak, *Magnetism in Medicine: A Handbook*, 2nd ed. (Weinheim: Wiley-VCH, 2007), 3–22.

16. John P. O'Reardon, Murat Altinay, and Pilar Cristancho, "Transcranial Magnetic Stimulation: A New Treatment Option for Major Depression," *Psychiatric Times* 27, no. 9 (2010).

17. In Mesmer's words,

> I believe I have grasped in nature the mechanism of influences, which [. . .] consist in a kind of reciprocal and alternative *flow* of the entering and outgoing currents, a subtle fluid, filling the space between two bodies. The need for this flow is founded on the law of space as full; i.e., in space filled with matter, it cannot be made a displacement without a replacement; which supposes that if a movement of the subtle matter is caused in a body, there is at once a similar movement in another likely to receive it, whatever the distance between the bodies. This kind of circulation is able to excite and reinforce in them properties analogue to their organization, which can be easily conceived by reflecting on *the continuity of the fluid matter*, and on its extreme mobility always equivalent to its subtlety: the magnet, electricity, as well as fire, offer to us models and examples of this universal law. (Franz Anton Mesmer, *Mémoire de F. A. Mesmer [. . .] sur ses découvertes* [Paris: Fuchs, 1799], 44–47; my emphasis)

18. Darnton, *End of the Enlightenment*, 14.

19. Geoffrey Sutton, "Electric Medicine and Mesmerism," *Isis*, no. 72 (1981): 83–84, 375–92. See also François Zanetti, "Magnétisme animal et électrcité médicale au dix-huitième siècle," in *Mesmer et mesmérismes: Le magnétisme animal en contexte*, ed. Bruno Belhoste and Nicole Edelman (Paris: Omniscience, 2015), 103–17.

20. Betsy Van Schlun, *Science and the Imagination: Mesmerism, Media, and the Mind in Nineteenth-Century English and American Literature* (Berlin: Galda + Wilch Verlag, 2007), 265.

21. Luigi Galvani, "De Viribus Electricitatis," in *Literature and Science in the Nineteenth Century: An Anthology*, ed. Laura Otis (Oxford: Oxford University Press, 2002), 136. Such displays would later inspire idealist imagery in Romantic poetry. Gilmore, *Aesthetic Materialism: Electricity and American Romanticism*, 70.

22. Franz Anton Mesmer and Louis Caullet de Veaumorel, *Aphorismes de M. Mesmer [. . .]* (Paris: M. Quinquet l'aîné, 1785), 131–36.

23. Bertrand Méheust calls "the Puységur event" the emergence of the unconscious as an object of study that would eventually revolutionize the notion of the self during the nineteenth century. See his preamble in Bertrand Méheust, *Somnambulisme et médiumnité, 1784–1930*, 2 vols. (Le Plessis-Robinson: Institut Synthélabo pour le progrès de la connaissance, 1999). See also Léon Chertok and Raymond de Saussure, *The Therapeutic Revolution, from Mesmer to Freud* (New York: Brunner/Mazel, 1979).

24. Before 1784, natural somnambulism usually referred to sleepwalking. For a study of the word *somnambulism*, see Nicole Edelman, *Voyantes, guérisseuses et visionnaires en France: 1785–1914* (Paris: A. Michel, 1995), 9.

25. For a detailed report on articles and theories treating the subject, see Jules Denis Du Potet, "Électro-Magnétisme," *Journal du*

magnétisme 12 (1853). For earlier reports, see previous volumes of the same journal.

26. Alison Winter, *Mesmerized: Powers of Mind in Victorian Britain* (Chicago: University of Chicago Press, 1998), 73, 276.

27. Michael Faraday, "On Table-Turning," in *Experimental Researches in Chemistry and Physics* (London: R. Taylor and W. Francis, 1859), 382.

28. For a recent study on animal magnetism in the United States, see Emily Ogden, *Credulity: A Cultural History of U.S. Mesmerism*, ed. Kathryn Lofton and John Lardas Modern, Class 200: New Studies in Religion (Chicago: University of Chicago Press, 2018).

29. The attractive power of magnetism may have played a role in the emergence of the legend of the sirens: "Legends of powerful mountains, islands or rocks capable of attracting boats and even pulling the nails from their hulls grew out of classical stories. They were elaborated in Arab folk tales, such as The Thousand and One Nights and discussed by medieval geographers and sailors. Italians suspected that the island of Elba disturbed their compasses. In the sixteenth century, the idea developed into a single landmass situated some distance from the North Pole. It controlled magnetic direction and created variation" (Stephen Pumfrey, *Latitude and the Magnetic Earth* [New York: MJF Books / Fine Communications, 2006], 84).

30. Poe, "The Spectacles," 890, 905. *La Sonnambula* played an integral part in bringing to the fore a cultural overlap between Victorian opera and mesmerism. Daniel Pick, *Svengali's Web: The Alien Enchanter in Modern Culture* (New Haven: Yale University Press, 2000), 8.

31. Joseph Philippe François Deleuze, *Practical Instruction in Animal Magnetism, Part 1*, trans. Thomas C. Hartshorn, 2nd ed. (Providence: B. Cranston & Company, 1837), 255.

32. Edgar Allan Poe, "Review of Human Magnetism," in *The Complete Works of Edgar Allan Poe*, ed. James Albert Harrison (New York: Thomas Y. Crowell & Company, 1845), accessed June 2019, https://www.eapoe.org/works/harrison/jah12c08.htm.

33. Chauncy Hare Townshend, *Facts in Mesmerism, with Reasons for a Dispassionate Inquiry into It* (London: Longman, Orme, Green, & Longman, 1840), 409, 84, 90.

34. Sebastian Matzner, *Rethinking Metonymy: Literary Theory and Poetic Practice from Pindar to Jakobson* (Oxford: Oxford University Press, 2016), 47–48.

35. Edgar Allan Poe, *Eureka*, ed. Stuart Levine and Susan F. Levine (Urbana: University of Illinois Press, 2004), 27, 36.

36. Edgar Allan Poe, "Mesmeric Revelation," in *The Works of the Late Edgar Allan Poe*, ed. Rufus Wilmot Griswold (New York: J. S. Redfield, Clinton Hall, 1844), 110, accessed June 2019, http://www.eapoe.org/works/TALES/mesmerd.htm.

37. Val Dusek, *The Holistic Inspirations of Physics: The Underground History of Electromagnetic Theory* (New Brunswick: Rutgers University Press, 1999), 36.

38. Martha Baldwin, "Athanasius Kircher and the Magnetic Philosophy" (Ph.D. diss., University of Chicago, 1987), 463.

39. M. de Bruno, *Recherches sur la direction du fluide magnétique* (Amsterdam: Chez Gueffier, 1785), 15–16.

40. This is the paragraph, which is not cited in Bachelard:

> Mais si, au lieu de s'abandonner à de vains systèmes formés sans le concours des expériences, on avoit étudié cette partie agissante de la nature, quel intéressant point de vue n'auroit-elle point offert au Physicien, au Naturaliste & au Philosophe. Quels rapports admirables n'auroit-on pas trouvés entre la manière d'être de tous les corps qui composent ce vaste univers! Rien ne prouve mieux la vérité de cette idée sublime de l'immortel Buffon, que la nature simple dans ses principes, ne parcourt la chaîne immense de tous les êtres, que par des nuances qui lient les uns aux autres. Il les a indiquées depuis les différentes classes des hommes, jusqu'aux dernières de celles des végétaux. Mais toutes ces classes à qui la vie sembleroit être une propriété exclusive,

seroient donc séparées du reste de la nature par un intervalle qui contrarieroit cette idée si sage et si vraie! (Ibid., 14–15)

41. Gaston Bachelard, *The Formation of the Scientific Mind: A Contribution to a Psychoanalysis of Objective Knowledge*, trans. Mary McAllester Jones (Manchester: Clinamen, 2002), 154–71.

42. Rare instances of asexual reproduction in insects (e.g., worms and aphids) helped persuade him to deem the polyp an animal. Abraham Trembley, *Mémoires pour servir à l'histoire d'un genre de polypes d'eau douce, à bras en forme de cornes* (Leide: J. & H. Verbeek, 1744), 1–20, 229–312.

43. Justinus Kerner, *The Seeress of Prevorst: Being Revelations Concerning the Inner-Life of Man, and the Inter-diffusion of a World of Spirits in the One We Inhabit*, trans. Catherine Crowe (London: J. C. Moore, 1845), 72.

44. Ibid., 7–10, 15, 19, 22.

45. Hegel took mesmerism seriously and wrote extensively on the subject. Glenn Alexander Magee, *Hegel and the Hermetic Tradition* (Ithaca: Cornell University Press, 2001), 215–21. See also Georg Wilhelm Friedrich Hegel, *Le magnétisme animal: Naissance de l'hypnose*, ed. François Roustang, Quadrige Grands Textes (Paris: Presses Universitaires de France, 2005). For a reading of Hegel's text on animal magnetism where Jean-Luc Nancy reimplements the notion of *partage* that he had developed in his analysis of *Ion*, see Jean-Luc Nancy, Mikkel Borch-Jacobsen, and Eric Michaud, *Hypnoses* (Paris: Editions Galilée, 1984).

46. Kerner, *Seeress of Prevorst*, 151–52.

47. See Derrida's reading of Rousseau in Jacques Derrida, *De la grammatologie* (Paris: Éditions de Minuit, 1997, 1997).

48. Kerner, *Seeress of Prevorst*, 31–36.

49. Saïd Hammoud, *Mesmérisme et romantisme allemand: 1766–1829* (Paris: L'Harmattan, 1994), 176.

50. Kerner, *Seeress of Prevorst*, 37.

51. Ibid., 40.

52. Ibid., 43, 51, 55.

53. Hammoud, *Mesmérisme et romantisme*, 186.

54. For more on haunted writing in the German context, see Avital Ronell, *Dictations: On Haunted Writing* (Urbana: University of Illinois Press, 2006). Laurence A. Rickels, *Aberrations of Mourning* (Minneapolis: University of Minnesota Press, 2011).

55. Kerner, *Seeress of Prevorst*, 80–81.

56. Nicolas Abraham, Maria Torok, and Jacques Derrida, *The Wolf Man's Magic Word: A Cryptonymy* (Minneapolis: University of Minnesota Press, 1986), 5.

57. Hauffe was her grandparents' favorite grandchild. She lived with them between the ages of five and twelve. During that time, evidence suggests that Hauffe went through a traumatic sexual experience. Wouter J. Hanegraaff, "A Woman Alone: The Beatification of Friederike Hauffe *née* Wanner (1801–1829)," in *Women and Miracle Stories*, ed. Anne-Marie Korte (Leiden: Brill, 2000), 221–25.

58. The biographical details on Kerner are from Hammoud, *Mesmérisme et romantisme*.

59. On the relation between the living and the dead, Kerner quotes Schelling at length: "Death, so far from weakening our personality, exalts it, since it frees it from so many contingencies. *Remembrance* is but a feeble expression to convey the intimate connexion which exists betwixt those who are departed and those who remain. In our innermost being we are in strict union with de dead; for in our better part we are no other than what they are—spirits. [. . .] when the external world sinks from us, the inner life ascends" (*Seeress of Prevorst*, 7). According to Kerner, Schelling wrote these lines in 1811, soon after the death of his wife.

60. Before *The Seeress of Prevorst* was translated into English in 1845, Poe had most likely read about Hauffe's extraordinary fits as early as 1843, when she was evoked at length in Margaret Fuller's *Summer on the Lakes* (1843). Mills, *Transition States*, 115–16.

61. Poe, "Mesmeric Revelation," 7. Subsequent quotations are also taken from this source.

62. Ibid., 120.

63. Kerner, *Seeress of Prevorst*, 333–34.

64. Edgar Allan Poe, "The Facts in the Case of M. Valdemar," in *Collected Works of Edgar*

Allan Poe: Tales and Sketches, 1843–1849, ed. Thomas Ollive Mabbott (Cambridge: Belknap Press of Harvard University Press, 1978), 1243, accessed June 2019, http://www.eapoe.org/works/tales/vldmara.htm.

65. For a detailed account of Poe's mesmeric hoaxes, see Antoine Faivre, "Borrowings and Misreading: Edgar Allan Poe's 'Mesmeric' Tales and the Strange Case of Their Reception," *Aries* 7 (2007).

66. Tresch also observes the appearance of a different kind of chain pattern in Poe's ironic and fragmentary tales: "We might extrapolate from these works that Poe was replacing the unbroken linearity of the classic chain of being—still operative in natural history and cosmology, though undergoing transfigurations—with an arabesque spatialization of expansive enclosures." John Tresch, "'Matter No More': Edgar Allan Poe and the Paradoxes of Materialism," *Critical Inquiry* 42, no. 4 (2016): 886.

67. The following passage from Thompson summarizes one of the important aspects of German "Romantic irony" that had a crucial influence on Poe's work:

> Around 1800, Friedrich Schlegel had conjoined the terms *irony* and *transcendentalism*. Irony was the process of transcending both the illusions of the world and the delusions of one's own limited mind. Such transcendence of the visible world and of the self was, for Schlegel, achieved through a sense of the comic and the absurd in the serious. By comparing successive phases of our own stupidity and shrewdness, Schlegel suggested, we evolve increasingly superior versions of the self, a true sense of irony. Such a true sense of irony is the perception or creation of a succession of contrasts between the ideal and the real, the serious and the comic, the sinister and the absurd, through which the "transcendental ego" can mock its own convictions and productions from the height of the "ideal." As Poe remarked in "The Philosophy of Composition" (1845), one wants in dark and fantastic composition to approach "as nearly to the ludicrous as [is] admissible" (H 14:205). This kind of higher and more objective skepticism can be seen, according to Friedrich Schlegel, in the works of Cervantes and Shakespeare, which are "full of artfully arranged confusion, charming symmetry of contrasts, marvelous alternation of enthusiasm and irony." (Gary Richard Thompson, *Poe's Fiction: Romantic Irony in the Gothic Tales* [Madison: University of Wisconsin Press, 1973], 17, 27)

68. Poe, "Mesmeric Revelation," 112.

69. Edgar Allan Poe, "Poe to James R. Lowell, July 2, 1844," in *The Letters of Edgar Allan Poe*, ed. John Ward Ostrom (New York: Gordian Press, 1966), accessed June 2019, https://www.eapoe.org/works/ostlttrs/pl661c06.htm#pg0258. Edgar Allan Poe, "Poe to Philip P. Cooke, August 9, 1846," in *The Collected Letters of Edgar Allan Poe*, ed. John Ward Ostrom, Burton R. Pollin, and Jeffrey A. Savoye (New York: Gordian Press, 2008), 595, accessed June 2019, https://www.eapoe.org/works/ostlttrs/pl081c08.htm.

70. Edgar Allan Poe, "The Murders in the Rue Morgue," in *Collected Works of Edgar Allan Poe: Tales and Sketches, 1831–1842*, ed. Thomas Ollive Mabbott (Cambridge: Belknap Press of Harvard University Press, 1978), 533, accessed June 2019, https://www.eapoe.org/works/mabbott/tom2t043.htm.

71. Dupin's analytical method derives its power of detection from the metonymic synergy of reason and the imagination; or, in Paul Hurh's words, "their interrelation through their opposition." Paul Hurh, "'The Creative and the Resolvent': The Origins of Poe's Analytical Method," *Nineteenth-Century Literature* 66, no. 4 (2012): 473. As discussed further in the next part, the magnet's opposite yet interrelated poles were a popular model to explore such puzzling metonymic connection during the Romantic era. The narrator's allusion to "the old philosophy of the Bi-Part Soul" refers to Aristotle's psychology in *Nicomachean Ethics* and its division between the rational and irrational parts of the soul. Stephanie Craighill, "The Influence of Duality and Poe's Notion of the 'Bi-Part Soul' on the Genesis of Detective

Fiction in the Nineteenth-Century" (2010), 39–40.

72. Poe, "Rue Morgue," 535.

73. Poe, "Poe to Philip P. Cooke," 595–96.

74. Poe, "Spectacles," 888. Unlike in Mabbott's version, Poe italicizes the French names in the only surviving manuscript. Joseph J. Moldenhauer, "Poe's 'The Spectacles': A New Text from Manuscript Edited, with Textual Commentary and Notes," *Studies in the American Renaissance* (June 1977): 200.

75. Poe, "Rue Morgue," 553.

76. Although Poe uses at times the term *deduction*, he mainly relies on *induction* to characterize Dupin's method of detection. Messac argues that for Poe the two methods were probably not fundamentally separate when, in a clairvoyant state, the detective could practice both simultaneously. Régis Messac, *Le "Detective novel" et l'influence de la pensée scientifique* (Paris: Honoré Champion, 1929), 34–38, 360–64.

77. See the first twenty aphorisms in book 2 of Bacon, *New Organon*.

78. Poe, *Eureka*, 10–11.

79. For a detailed study on Poe's appropriation of "inductive reasoning," see Loisa Nygaard, "Winning the Game: Inductive Reasoning in Poe's 'Murders in the Rue Morgue,'" *Studies in Romanticism* 33, no. 2 (1994).

80. Umberto Eco et al., *The Sign of Three: Dupin, Holmes, Peirce*, ed. Umberto Eco and Thomas A. Sebeok, Advances in Semiotics (Bloomington: Indiana University Press, 1983); Paul Grimstad, "C. Auguste Dupin and Charles S. Peirce: An Abductive Affinity," *Edgar Allan Poe Review* 6, no. 2 (2005); Axel Gelfert, "Observation, Inference, and Imagination: Elements of Edgar Allan Poe's Philosophy of Science," *Science and Education* 23, no. 3 (2014); Ilkka Niiniluoto, *Truth-Seeking by Abduction* (Cham, Switzerland: Springer, 2018). For an overview of Peirce's changing conception of abduction, see Sami Paavola, "Peircean Abduction: Instinct or Inference?," *Semiotica*, no. 153–1/4 (2005).

81. Another study by a philosopher of science that dismisses Poe's appropriation of the term *induction* and, in turn, loses track of its important allusion to electromagnetism is David N. Stamos, *Edgar Allan Poe, Eureka, and Scientific Imagination* (Albany: State University of New York Press, 2017), 416–17, 40–47.

82. Charles J. Rzepka, *Detective Fiction* (Cambridge: Polity, 2005), 45, 80–81, 159–60.

83. Andrea Goulet, *Legacies of the Rue Morgue: Science, Space, and Crime Fiction in France* (Philadelphia: University of Pennsylvania Press, 2016), 4, 11.

84. Poe has been experimenting with narratives that could prompt readers to form a train of ratiocination at least since his 1835 tale "Berenice." Susan Elizabeth Sweeney, "Solving Mysteries in Poe, or Trying To," in *The Oxford Handbook of Edgar Allan Poe*, ed. J. Gerald Kennedy and Scott Peeples (Oxford: Oxford University Press, 2019), 194–95.

Chapter 2

1. "Comme la chute de la poire [qui] devint la cause première des découvertes de Newton." My translation. Honoré de Balzac, *Nouveaux contes philosophiques, par M. de Balzac. Maître Cornélius; Madame Firmiani; L'Auberge rouge; Louis Lambert* (Paris: Charles Gosselin, 1832), 333.

2. "Comme la sensation électrique toujours ressentie par Mesmer à l'approche d'un valet [qui] fut l'origine de ses découvertes en magnétisme." My translation. Honoré de Balzac, *Histoire intellectuelle de Louis Lambert, par M. de Balzac, fragment extrait des "Romans et contes philosophiques"* (Paris: Charles Gosselin, 1833), 111.

3. See chapters 3–5 in Eric Wilson, *Emerson's Sublime Science* (New York: St. Martin's Press, 1999).

4. Sean Ross Meehan, "Ecology and Imagination: Emerson, Thoreau, and the Nature of Metonymy," *Criticism* 55, no. 2 (2013): 308.

5. Laura Dassow Walls, "'Every Truth Tends to Become a Power': Emerson, Faraday, and the Minding of Matter," in *Emerson for the Twenty-First Century: Global*

Perspectives on an American Icon, ed. Barry Tharaud (Newark: University of Delaware Press, 2010), 301.

6. Albert Einstein, "Excerpt from Essay by Einstein on 'Happiest Thought' in his Life," *New York Times*, March 28, 1972. This excerpt is from an unpublished essay entitled "The Fundamental Idea of General Relativity in its Original Form," written about 1919.

7. Honoré de Balzac, "Avant-propos," in *La comédie humaine, Bibliothèque de la pléiade, vol. 1* (Paris: Gallimard, 1976), 11–12.

8. See the preface in Tresch, *The Romantic Machine: Utopian Science and Technology after Napoleon* (Chicago: University of Chicago Press, 2012). See also John Tresch, "Electromagnetic Alchemy in Balzac's The Quest for the Absolute," in *The Shape of Experiment*, ed. Henning Schmidgen and Julia Kursell (Berlin: Max-Planck preprint, 2007).

9. Sydney Lévy, "Balzac et la machine-*peau de chagrin*," *Théorie, littérature, enseignement* 21 (2003).

10. Balzac, "Avant-propos," 16–17.

11. Honoré de Balzac, *Lettres à l'étrangère* (Paris: Calmann-Lévy, 1906), 366.

12. William Whewell, *Of Induction: With Especial Reference to Mr. J. Stuart Mill's System of Logic* (London: John W. Parker, 1849), 76.

13. Cognitive linguists have recently begun to argue that metonymic reasoning is closer to induction and abduction than deduction. Klaus-Uwe Panther and Linda L. Thornburg, "What Kind of Reasoning Mode Is Metonymy?," in *Conceptual Metonymy: Methodological, Theoretical, and Descriptive Issues*, ed. Olga Blanco-Carrión, Antonio Barcelona, and Rossella Pannain (Amsterdam: John Benjamins, 2018).

14. Honoré de Balzac, *Le père Goriot* (Paris: Le livre de poche, 1983), 227. All translations (with modifications) taken from http://www.gutenberg.org/.

15. Ibid., 204, 21–22.

16. Honoré de Balzac, *Louis Lambert*, in La comédie humaine, Bibliothèque de la pléiade, vol. 10 (Paris: Gallimard, 1935), 865.

17. Dahlia Porter, *Science, Form, and the Problem of Induction in British Romanticism*, ed. James Chandler, Cambridge Studies in Romanticism (Cambridge: Cambridge University Press, 2018).

18. Balzac, "Avant-propos."

19. Ibid., 17.

20. Ibid., 1136. Animal magnetism might have already been a strong influence on the theories of physiognomy. In 1785, after hearing about animal magnetism, J. C. Lavater "went to Geneva, where he received practical instruction in Puységurian methods." On his return, he reportedly used magnetism to cure his wife, who had been for a long time hopelessly ill. Alan Gauld, *A History of Hypnotism* (Cambridge: Cambridge University Press, 1992), 76–77.

21. In a letter to Zulma Carraud dating from 1833, Balzac writes, "Ma sœur a été guérie de la même maladie qu'a Mme Nivet, par une suite de magnétisme, par la simple action, répétée deux heures tous les jours, de ma mère. C'est un fait irrécusable." Honoré de Balzac, *Correspondance*, ed. Roger Pierrot, vol. 2 (Paris: Éditions Garnier frères, 1960), 312. Balzac's mother played an important role in shaping his belief in magnetism, mysticism, and religion.

22. Théophile Gautier, *Portraits contemporains*, in *Oeuvres complètes* (Genève: Slatkine Reprints, 1978), 73–74.

23. Nicole Edelman, *Voyantes, guérisseuses et visionnaires en France: 1785–1914* (Paris: A. Michel, 1995), 19.

24. According to Edelman, most somnambulists were women and most of the magnetizers were men. Balzac's somnambulists tend to be females or androgynous characters like Séraphîta. Louis Lambert is also described throughout the novel with feminine traits. For Mesmer there was no difference whether the magnetization was done by a male or a female. Franklin Rausky, *Mesmer: Ou, la révolution thérapeutique* (Paris: Payot, 1977), 226–27.

25. Edelman, *Voyantes*, 29.

26. Edelman writes, "Victor Race, le somnambule de Puységur, avait écrit ses prescriptions médicales. D'autre somnambules,

magnétisées par le Chevalier de Barberin écriront aussi." Ibid., 227. Although she is the best example among the many scientifically inclined somnambulist-writers mentioned in her book, Edelman could not verify the exact state in which Marie-Louise wrote these pamphlets.

27. Honoré de Balzac, *Ursule Mirouët*, in La comédie humaine, Bibliothèque de la pléiade, vol. 3 (Paris: Gallimard, 1935), 357–58.

28. Ibid., 280.

29. Ibid., 324.

30. Ibid.

31. Göran Blix, "The Occult Roots of Realism: Balzac, Mesmer, and Second Sight," *Studies in Eighteenth-Century Culture* 36 (2007): 263, 73, 76–77.

32. Ibid., 268.

33. Ibid., 277.

34. Honoré de Balzac, *Séraphîta*, in La comédie humaine, Bibliothèque de la pléiade, vol. 11 (Paris: Gallimard, 1976), 737.

35. Ibid., 823.

36. Christine Blondel, *A.-M. Ampère et la création de l'électrodynamique, 1820–1827* (Paris: Bibliothèque nationale, 1982), 69–70. For an up-to-date and detailed analysis of Ampère's instrumental role in the early development of electromagnetism and electrodynamics, see also Christine Blondel and Bertrand Wolff, "@. Ampère et l'histoire de l'électricité," accessed June 2019, http://www.ampere.cnrs.fr/.

37. According to his correspondence, Ampère was curious about animal magnetism and even witnessed Puységur and his somnambulist in action (letters to Bredin, April 5, 10, 28, and May 5, 1813). He also underwent what looks like a magnetic treatment that cured a bad case of sore throat (letter to Bredin, April 17, 1821). Ampère's correspondence can be found at Blondel and Wolff, "@. Ampère et l'histoire de l'électricité."

38. Honoré de Balzac, *Théorie de la démarche*, in La comédie humaine, Bibliothèque de la pléiade, vol. 12 (Paris: Gallimard, 1981), 269–70.

39. Balzac had probably seen Ampère in person. Madeleine Ambrière, *Balzac et la recherche de l'absolu* (Paris: Hachette, 1968), 178–80.

40. Honoré de Balzac, *La recherche de l'absolu*, Folio Classique (Gallimard, 2005), 119.

41. It should be noted that "*électro-magnétisme*" predates the discovery of electromagnetism (see the definition at http://www.cnrtl.fr). For example, "Les analogies sont d'un grand secours dans ces recherches, mais il faut s'en défier quelquefois, & toujours en user avec beaucoup de réserve. Le *pneumatisme*, l'*électricité*, le *magnétisme*, l'*électro-magnétisme* (peut-être) sont déjà les mots sacramentaux presqu'universels, & feront bientôt les notions fondamentales & essentielles, de la Physique moderne devenue toute chymique." Thouvenel, *Mémoire physique et médicinal montrant des rapports évidens entre les phénomènes de la baguette divinatoire, du magnétisme et de l'électricité* (Paris et Londres: Didot le jeune, 1781), 85. Balzac's usage has registered the new implications of the word, since for him, electromagnetism generates a "product" (phosphorus, and arguably movement or animation).

42. Michael Faraday, "On the Induction of Electric Currents [. . .]," in *Experimental Researches in Electricity* (London: R. and J. E. Taylor, 1839–55), 1.

43. John L. Heilbron, *Electricity in the 17th and 18th Centuries: A Study of Early Modern Physics* (Berkeley: University of California Press, 1979).

44. Tiberius Cavallo, *A Complete Treatise of Electricity in Theory and Practice; With Original Experiments* (London: Edward and Charles Dilly, 1777), 382. I could not find an eighteenth-century example that clearly signals the shift from old electrical terminologies to the verb *to induce*.

45. The Latin etymology of the verb *to induce* means "to lead." Samuel Johnson, "A Dictionary of the English Language" (London: J. F. and C. Rivington, 1785), https://archive.org/details/dictionaryofengl01johnuoft.

46. George John Singer, *Elements of Electricity and Electro-chemistry* (London: Longman, Hurst, Rees, Orme, Brown, and R. Triphook, 1814), 130.

47. Humphry Davy, *Elements of Chemical Philosophy*, vol. 1 (Philadelphia: Bradford and Inskeep, 1812), 74. As noted by Singer, in Davy's published works the apparition of the term *induction* for various electrical effects dates back at least to 1807. George John Singer, "Remarks on Some Electrical and Electrochemical Phenomena," *Journal of Natural Philosophy, Chemistry and the Arts* 31 (1812): 219.

48. For Singer, Davy conflated two fundamentally different types of electrical effects—namely, the redistribution and the communication of charges that an electrically charged object could provoke in a nearby conductor. Singer, "Remarks on Some Electrical and Electrochemical Phenomena," 217–19.

49. Anonymous, "Notices Respecting New Books," *Philosophical Magazine* 40 (1812): 435.

50. "Amid such varying adaptations of the word *induction* there is much to gain in allotting to the electrostatic induction of charges by charges the distinguishing name of *influence*, as suggested by Priestley." Silvanus P. Thompson, *Michael Faraday: His Life and Work* (London: Cassell, 1898), 153.

51. André-Marie Ampère, *Théorie des phénomènes électro-dynamiques, uniquement déduite de l'expérience* (Paris: Méquignon-Marvi, 1826).

52. Kenneth L. Caneva, "'Discovery' as a Site for the Collective Construction of Scientific Knowledge," *Historical Studies in the Physical and Biological Sciences* 35, no. 2 (2005): 184–88.

53. Faraday, "On the Induction of Electric Currents [. . .]," 16. For more details on Faraday's electromagnetic terminology, see Aaron D. Cobb, "History and Scientific Practice in the Construction of an Adequate Philosophy of Science: Revisiting a Whewell/Mill Debate," *Studies in History and Philosophy of Science* 42, no. 1 (2011): 90.

54. Friedrich Steinle, "Work, Finish, Publish? The Formation of the Second Series of Faraday's *Experimental Researches in Electricity*," *Physis* 33 (1996): 152–53.

55. Michael Faraday, "On the Induction of Electric Currents [. . .]," in *Experimental Researches in Electricity* [1839–55], 32; Faraday, "Terrestrial Magneto-electric Induction [. . .]," in *Experimental Researches in Electricity*, [1839–55], 66–67.

56. James Clerk Maxwell, *A Treatise on Electricity and Magnetism*, vol. 2 (Oxford: Clarendon Press, 1873), 162–64.

57. As mentioned in the introduction, Tresch has shown that Ampère's Newtonian works in electrodynamics were not detached from his romantic interests. Newton himself practiced an inductive method that was informed by his theological and historical concerns. Markley, *Fallen Languages*, 141–43.

58. James Clerk Maxwell, *A Treatise on Electricity and Magnetism*, vol. 2 (Oxford: Clarendon Press, 1873), 162–64.

59. William Whewell, "Modern Science—Inductive Philosophy," review of J. Hershel's *Preliminary Discourse on the Study of Natural Philosophy*, *Quarterly Review* 45 (1831): 379.

60. Cited in Laura J. Snyder, *Reforming Philosophy: A Victorian Debate on Science and Society* (Chicago: University of Chicago Press, 2006), 24.

61. Ibid., 100.

62. William Whewell, *Of Induction: With Especial Reference to Mr. J. Stuart Mill's System of Logic* (London: John W. Parker, 1849), 76.

63. John Stuart Mill, *A System of Logic, Ratiocinative and Inductive, Being a Connected View of the Principles of Evidence and the Methods of Scientific Investigation*, vol. 1 (London: John W. Parker, 1843), 456.

64. Ibid., 485–90.

65. Whewell, *Induction*, 48–50.

66. Pierre de Maricourt, *Letter of Petrus Peregrinus on the Magnet, A.D. 1269*, trans. Brother Arnold (New York: McGraw, 1269, 1904).

67. William Whewell, *The Philosophy of the Inductive Sciences: Founded Upon Their History*, vol. 1 (London: John W. Parker, 1840), 331–60.

68. Cobb, "History and Scientific Practice."

69. James Clerk Maxwell, *A Treatise on Electricity and Magnetism*, vol. 2 (Oxford: Clarendon Press, 1873), 162–64.

70. Steinle, "Work, Finish, Publish?," 144.

71. As Steinle notes,

> Far from being a mindless playing around with an apparatus, exploratory experimentation may well be characterized by definite guidelines and epistemic goals. The most prominent characteristic of the experimental procedure is the systematic variation of experimental parameters. The first aim here is to find out which of them are essential. Closely connected, there is the central goal of formulating empirical regularities about these dependencies and correlations. Typically they have the form of "if—then" propositions, where both the if- and the then- clauses refer to the empirical level. In many cases, however, the attempt to reformulate regularities requires the revision of existing concepts and categories, and the formation of new ones, which allow a stable and general formulation of the experimental results. It is here, in the realm of concept-formation, where exploratory experimentation has its most unique power and importance. There is, finally, often the attempt to develop experimental arrangements that involve only the necessary conditions for the effect in question and thus represent the general regularity or law in a most obvious way. Those experiments are attributed a particular status in that they serve as core effects to which all other phenomena of the field can be "reduced." (Friedrich Steinle, "Experiments in History and Philosophy of Science," *Perspectives on Science* 10, no. 4 [2002]: 419)

See also Friedrich Steinle, *Exploratory Experiments: Ampère, Faraday, and the Origins of Electrodynamics*, trans. Alex Levine (Pittsburgh: University of Pittsburgh Press, 2016).

72. For a theoretical exploration of this struggle in the works of Maxwell and Hendrik Lorentz (1853–1928), see the chapter on "the ether and the opposition between the continuous/discontinuous" in Françoise Balibar, *Einstein 1905: De l'éther aux quanta* (Paris: Presses universitaires de France, 1992), 54–89.

73. Bruce J. Hunt, *Pursuing Power and Light: Technology and Physics from James Watt to Albert Einstein* (Baltimore: Johns Hopkins University Press, 2010), 142–48.

74. Albert Einstein, "On the Electrodynamics of Moving Bodies," in *The Swiss Years: Writings, 1900–1909*, English Translation Supplement, The Collected Papers of Albert Einstein (Princeton: Princeton University Press, 1989), 140.

75. Albert Einstein, "Field Theories, Old and New," *New York Times*, February 3, 1929.

76. Albert Einstein, "Excerpt from Essay by Einstein on 'Happiest Thought' in His Life," *New York Times*, March 28, 1972.

77. Ibid.

Chapter 3

1. Sam Halliday, *Science and Technology in the Age of Hawthorne, Melville, Twain, and James: Thinking and Writing Electricity* (New York: Palgrave Macmillan, 2007), 99, 111.

2. "Ce n'est pas pour moi que je vous dis cela; c'est pour vous-même et les autres, afin que le prestige du secret retienne dans les limites du devoir et de la vertu ceux qui, aimantés par l'électricité de l'inconnu, seraient tentés de m'imiter." Lautréamont, *Les chants de Maldoror* (Paris: Flammarion, 1990), 266.

3. "Le Poète se fait *voyant* par un long, immense et raisonné *dérèglement* de *tous les sens*." Arthur Rimbaud, *Œuvres*, ed. Suzanne Bernard and André Guyaux, Classiques Garnier (Paris: Garnier frères, 1983), 348–49. Translations from Arthur Rimbaud, Wallace Fowlie, and Seth Adam Whidden, *Rimbaud: Complete Works, Selected Letters: A Bilingual Edition*, trans. Wallace Fowlie (Chicago: University of Chicago Press, 2005), 377, 79.

4. Rimbaud, *Œuvres*, 348–49. Translation from Rimbaud, Fowlie, and Whidden, *Rimbaud*, 117.

5. The expression *electric butterflies* refers to electrostatics experiments known since the eighteenth century that produced sparks fluttering like butterflies. Jean-Sébastien-Eugène Julia De Fontenelle and François Malepeyre, *Nouveau manuel complet de physique amusante ou nouvelles récréations physiques* (Paris: Roret, 1860), 146.

6. Rimbaud was not the only important poet inspired by the telegraph. For other examples, see Jason R. Rudy, *Electric Meters: Victorian Physiological Poetics* (Athens: Ohio University Press, 2009).

7. See, for instance, Sharalyn Orbaugh, "Emotional Infectivity: Cyborg Affect and the Limits of the Human," in *Mechademia 3: Limits of the Human*, ed. Frenchy Lunning (Minneapolis: University of Minnesota Press, 2008); Mary Ann Doane, "Technophilia: Technology, Representation, and the Feminine," in *The Gendered Cyborg: A Reader*, ed. Gill Kirkup et al. (London: Routledge, 2000). Gaby Wood, *Edison's Eve: A Magical History of the Quest for Mechanical Life* (New York: Anchor Books, 2002); Martin Willis, *Mesmerists, Monsters, and Machines: Science Fiction and the Cultures of Science in the Nineteenth Century* (Kent: Kent State University Press, 2006); Carol de Dobay Rifelj, "Minds, Computers and Hadaly," in *Jeering Dreamers: Villiers de l'Isle-Adam's L'Eve Future at Our Fin de Siècle: A Collection of Essays*, ed. John Anzalone (Amsterdam: Rodopi, 1996); Jacques Noiray, *Jules Verne, Villiers de L'Isle-Adam*, vol. 2, Le Romancier et la machine (Paris: Corti, 1982).

8. See chapters 5 and 6 in Minsoo Kang, *Sublime Dreams of Living Machines: The Automaton in the European Imagination* (Cambridge: Harvard University Press, 2011).

9. The discovery of electromagnetism played an integral part in development of the concept of energy. During the second half of the eighteenth century, the steam engine began to power the Industrial Revolution by converting heat into a mechanical force. In 1800 the study of animal electricity led to Alessandro Volta's invention of the battery, which showed that the force of chemical affinity could produce an electric current. This electric current could in turn generate heat, make a wire glow, and trigger another chemical reaction, as in the electrolysis of water. What electromagnetism demonstrated is that an electric current behaved like a magnet and that due to induction, a mechanical force could produce an electric current, or vice versa. The interconvertibility of mechanical force, heat, chemical affinity, light, electricity, and magnetism implied that these wide-ranging phenomena were in fact equivalent to a grand unifying entity called "energy." Conversion processes also provided the empirical ground for the law of conservation of energy (a.k.a. the first law of thermodynamics) by confirming that in a closed system, energy can change form but is neither created nor destroyed. See the experiments on the mechanical power of electromagnetism in James Prescott Joule, *The Scientific Papers of James Prescott Joule*, 2 vols. (London: Taylor and Francis, 1884). See also Thomas Kuhn, "Energy Conservation as an Example of Simultaneous Discovery," in *The Essential Tension: Selected Studies in Scientific Tradition and Change* (Chicago: University of Chicago Press, 1959). And Bruce J. Hunt, *Pursuing Power and Light: Technology and Physics from James Watt to Albert Einstein* (Baltimore: Johns Hopkins University Press, 2010), 25–40.

10. Fumaroli calls for such a shift in his brief paper "Histoire de la littérature et histoire de l'électricité: Actes du colloque de l'Association pour l'histoire de l'électricité en France, Paris, 11–13 octobre 1983," in *L'Électricité dans l'histoire, Problèmes et méthodes* (Paris: Presses universitaires de France, 1985).

11. Pierre Juhel, *Histoire de la boussole: L'aventure de l'aiguille aimantée* (Versailles: Editions Quae, 2013).

12. Cited in Stephen Pumfrey, *Latitude and the Magnetic Earth* (New York: MJF Books / Fine Communications, 2006), 49.

13. Pierre de Maricourt, *Letter of Petrus Peregrinus on the Magnet, A.D. 1269*, trans. Brother Arnold (1269; repr., New York: McGraw, 1904), 14.

14. Ibid., 16.

15. Ibid., 32–33.

16. To this day, scientists struggle to predict the geomagnetic fluctuations manifested by these irregular phenomena called *magnetic variation* and *inclination*. The Earth is "a magnetic system of extraordinary complexity that is still little understood. The modern

consensus is that the Earth's rotation sets up currents in the molten core, and creates an electromagnetic dynamo. The currents are arranged more or less symmetrically around the Earth's axis of rotation. [. . .] Dynamo theorists invoke a number of factors, some secure and others speculative, to explain why the Earth's field is most like a tilted dipole." Pumfrey, *Latitude and the Magnetic Earth*, 40.

17. Aphorisms 54 and 129 in Francis Bacon, *The New Organon*, ed. Lisa Jardine and Michael Silverthorne (Cambridge: Cambridge University Press, 2000).

18. "Thus Aristotle's world would seem to be a monstrous creation, in which all things are perfect, vigorous, animate, while the earth alone, luckless small fraction, is imperfect, dead, inanimate, and subject to decay. On the other hand, Hermes, Zoroaster, Orpheus, recognize a universal soul. As for us, we deem the whole world animate, and all the globes, all stars, and this glorious earth, too, we hold to be from the beginning by their own destinate souls governed." William Gilbert, *On the Magnet* (New York: Basic Books, 1958), 309.

19. Ibid., 98, 110. Since antiquity, the mysterious attraction between magnets has provided a convenient way to talk about the equally mysterious attraction of love. For lack of a better explanation, ancient thinkers and poets also described the attraction of iron toward the magnet in Romantic terms. Richard Wallace, "'Amaze Your Friends!' Lucretius on Magnets," *Greece and Rome* 43, no. 2 (October 1996): 183.

20. Aristotle, "On the Soul," accessed June 2019, http://classics.mit.edu/Aristotle/soul.1.i.html.

21. This quote is the title of the final chapter of book 5 in Gilbert, *On the Magnet*.

22. Ibid., 1, 8.

23. See aphorism 54 and 64 Bacon, *New Organon*.

24. Although *magnesia* can mean different things in alchemical lore, it is often associated with the lodestone. See Geoffrey Chaucer, "The Canon's Yeoman's Tale," in *Canterbury Tales*, ed. Gerard NeCastro, lines 1448–71, accessed June 2019, https://medievalit.com/home/echaucer/original-texts/the-canons-yeomans-tale-original-text/.

25. For Chaucer's and Bacon's ambiguous relation to alchemy, see Stanton J. Linden, *Darke Hierogliphicks: Alchemy in English Literature from Chaucer to the Restoration* (Lexington: University Press of Kentucky, 1996).

26. This passage is from a letter cited in Dusek, *The Holistic Inspirations of Physics: The Underground History of Electromagnetic Theory*, 169. On Gilbert's influence on Kepler, see also Pumfrey, *Latitude and the Magnetic Earth*, 160, 216–21.

27. In 1623 Kepler sums up the transition from an animist to a mechanical cosmology in a footnote in his *The Mystery of the Universe*: "If the word soul (*anima*) be replaced by force (*vis*), we have the very principle on which the celestial physics is based." Cited in Dusek, *Holistic Inspirations of Physics*, 167.

28. My understanding of magnetism in the eighteenth century is particularly indebted to Fara, *Sympathetic Attractions: Magnetic Practices, Beliefs, and Symbolism in Eighteenth-Century England.* See also Patricia Fara, *Fatal Attraction: Magnetic Mysteries of the Enlightenment* (New York: MJF Books / Fine Communications, 2006).

29. Plate XXII in René Descartes, *Principles of Philosophy*, trans. Valentine Rodger Miller and Reese P. Miller (Dordrecht: D. Reidel, 1983).

30. For the polemic between Newtonians and Cartesians, see Alexandre Koyré, *Du monde clos à l'univers infini* (Paris: Gallimard, 2005). I am referring to the fourteenth letter in Voltaire, *Lettres philosophiques* (Paris: Flammarion, 2006), 146–51. See also Dusek, *Holistic Inspirations of Physics*, 185.

31. Edmond Halley, "A Discourse Concerning Gravity, and Its Properties, Wherein the Descent of Heavy Bodies, and the Motion of Projects Is Briefly, but Fully Handled:

Together with the Solution of a Problem of Great Use in Gunnery," *Philosophical Transactions*, no. 16 (1686): 5.

32. For details on Halley's influence on subterranean voyage literature, which includes Poe, see Fara, *Fatal Attraction*, 62–68.

33. Halley's initial intuition on the magnetic nature of the aurora borealis proved to be partly true. The aurora borealis occurs due to an interaction between the solar wind and geomagnetism. Edmond Halley, "An Account of the Late Surprizing Appearance of the Lights Seen in the Air, on the Sixth of March Last; with an Attempt to Explain the Principal Phaenomena Thereof; as It Was Laid Before the Royal Society," *Philosophical Transactions* 29, no. 347 (1716): 423.

34. Benjamin Franklin, *Memoirs of Benjamin Franklin*, ed. William Temple Franklin et al., 2 vols. (Philadelphia: M'Carty & Davis, 1834), 326. Around the same time, Jan Hendrik van Swinden also opposed this "analogy." Jan Hendrik van Swinden, *Recueils de mémoires sur l'analogie de l'électricité et du magnétisme*, 3 vols. (La Haye: Les libraires associés, 1784).

35. Cited in Christine Blondel, *A.-M. Ampère et la création de l'électrodynamique, 1820–1827* (Paris: Bibliothèque nationale, 1982), 19.

36. Faraday thought that his idea of lines of force could provide an alternative to Newtonian action at a distance. Although Newton himself did not believe that gravity could operate in a vacuum, his law of gravitational attraction ($F = Gm1m2/r^2$) implied that its action took place between two masses (m1 and m2) instantaneously through space as if without mediation. Faraday criticized such action at a distance with a thought experiment:

> The notion of the gravitating force is, with those who admit Newton's law, but go with him no further, that matter attracts matter with a strength which is inversely as the square of the distance. Consider, then, a mass of matter (or a particle), for which present purpose the sun will serve, and consider a globe like one of the planets, as our earth, either created or taken from distant space and placed near the sun as our earth is;—the attraction of gravity is then exerted, and we say that the sun attracts the earth, and also that the earth attracts the sun. But if the sun attracts the earth, that force of attraction must either arise *because* of the presence of the earth near the sun; or it must have *pre-existed* in the sun when the earth was not there. If we consider the first case, I think it will be exceedingly difficult to conceive that the sudden presence of the earth, 95 millions of miles from the sun, and having no previous physical connexion with it, nor any physical connexion caused by the mere circumstance of juxtaposition, should be able to raise up in the sun a power having no previous existence. (Faraday, "On Some Points of Magnetic Philosophy," 571–72)

The physics behind action at a distance appear to violate the law of the conservation of force. One way to solve this problem is to consider the sun's gravitational attraction as a kind of potential force readily available throughout space before Earth's arrival. Attraction extends then beyond the sun in the form of a potential for action that permeates the solar system. Such a preexisting gravitational/spatial condition is what the earth detects and renders manifest when it is attracted by the sun. Faraday brings forth an alternative model to Newtonian action at a distance by considering it instead a "case of a constant necessary condition to action in space" (ibid., 574). Whereas space remains unaltered or absolute in Newtonian attraction, Faraday's notion of a "constant necessary condition to action in space" (or what is known today as a *field* or as *action by continuous contact*) implied that space played an integral part in the interaction between distant bodies and that it was malleable. The presence of a body in space modified the configuration of this space, and such modifications were responsible for attraction. Faraday made the physical configuration of such fields visible through his work on electric (and more specifically, magnetic) lines of force, which were mathematically useful because they displayed a direction and way to quantify the magnitude of the force anywhere in the field.

37. *Naturphilosophie* is a vague historical classification of heterogeneous ideas. My usage of this term focuses on Schelling's and Goethe's notion of polarity. On the difficulty to define *Naturphilosophie* and the related expression *Romantic science* see Caneva, "Physics and Naturphilosophie: A Reconnaissance," *History of Science* 35, no. 107 (1997); Friedrich Steinle, "Romantic Experiment? The Case of Electricity" (paper presented at the Ciencia y romanticismo, Maspalomas, 2002).

38. Friedrich Wilhelm Joseph von Schelling, *First Outline of a System of the Philosophy of Nature*, trans. Keith R. Peterson (Albany: State University of New York Press, 2004).

39. "Self-contained world system" refers to Schelling's concept of nature, where "Nature is absolute identity within itself—absolutely equal to itself—and yet in this identity it is opposed to itself once more, object to itself.—The general expression of Nature is therefore 'identity in duplicity and duplicity in identity.'" Nature becomes "object to itself" in, for example, the scientific works of natural philosophers, who are themselves an integral part of nature and thus cannot apprehend it from a hypothetical site outside of nature. Nature and the study of nature, then, present an "inner contradiction," which becomes tenable only through the inorganic, or primal, phenomenon of magnetism. According to Schelling, "This inner contradiction cannot be known originally, it is known only in the *phenomenon of magnetism*; in the latter alone do we distinguish universal duplicity in its *first* origin" (ibid., 180).

40. Ibid., 181. This quote is from a series of additional handwritten notes to his 1799 *First Outline of a System of the Philosophy of Nature*.

41. Jeremy Adler, "The Aesthetics of Magnetism: Science, Philosophy, and Poetry in the Dialogue Between Goethe and Schelling," in *The Third Culture: Literature and Science*, ed. Elinor S. Shaffer, European Cultures (Berlin: Walter de Gruyter, 1998).

42. Adler's translation. Ibid., 91. Val Dusek has argued that such "holistic" or "organic" ideas expressed in folk wisdom and *Naturphilosophie* anticipated and, for many during the nineteenth century, were confirmed by the discovery of electromagnetism. These ideas survived the rise of modern mechanical thinking and atomism due to marginalized groups such as women (particularly in the figures of the healer and witch) and the peasantry. The cultural and power transitions marked by the Renaissance and the French Revolution witnessed renewed interests in holistic, alchemical, and occult thinking. During the Renaissance, Paracelsus, in his attempt to reform medieval medical practices, recommended that his pupils learn from peasants and female healers. In 1600 Gilbert had associated his magnetic philosophy with the alchemical tradition, especially with its idea of a "universal soul," to distance himself from the then-dominant worldview of scholasticism. Closer to Schelling and Goethe, the French Revolution brought folk culture back to the fore, particularly through the writings of déclassé intellectuals and, later, by bohemians who functioned as witnesses to this predominantly illiterate part of society. Dusek, *Holistic Inspirations of Physics*.

43. Adler, "Aesthetics of Magnetism." 92.

44. Juhel, *Histoire de la boussole*, 83–115.

45. Mary Wollstonecraft Shelley, *Frankenstein*, ed. J. Paul Hunter, Norton Critical Edition (New York: W. W. Norton, 1996), 7–8.

46. Ibid., 30, 32.

47. Ibid., 34.

48. Marilyn Butler, "'Frankenstein' and Radical Science," in *Frankenstein*, ed. J. Paul Hunter (New York: W. W. Norton, 1996).

49. The invention of the electric battery also inspired attempts to develop a magnetic battery at the beginning of the nineteenth century. Robert de Andrade Martins, "Ørsted, Ritter, and Magnetochemistry," in *Hans Christian Ørsted and the Romantic*

Legacy in Science Ideas, Disciplines, Practices, ed. Robert Michael Brain, R. S. Cohen, and Ole Knudsen (Dordrecht: Springer, 2007), 344.

50. Shelley, *Frankenstein*, 155.

51. Thompson refers to John Cleves Symmes's 1820 *Symzonia: A Voyage of Discovery* as the main source for Reynolds's theory of a hollow earth. Edgar Allan Poe, *The Selected Writings of Edgar Allan Poe: Authoritative Texts, Backgrounds and Contexts, Criticism*, ed. Gary Richard Thompson (W. W. Norton, 2004), 430–31.

52. Edgar Allan Poe, "The Narrative of Arthur Gordon Pym. Of Nantucket," In *The Collected Writings of Edgar Allan Poe*, ed. Burton R. Pollin (New York: Gordian Press, 1994), 205–6.

53. "L'être vivant, dit-on, naît toujours d'un être semblable à lui. Mais, avant la découverte de l'aimantation électrique, on aurait pu dire également que tout aimant avait pour origine un aimant préexistant." Jean Rostand, *La vie et ses problèmes* (Paris: Flammarion, 1939), 14.

54. Oliver Lodge, *Life and Matter: A Criticism of Professor Haeckel's "Riddle of the Universe"* (New York: G. P. Putnam's Sons, 1905), 11.

55. Ibid., 123.

56. Ibid., 124–29.

57. Cited and translated in Koen Vermeir, "Athanasius Kircher's Magical Instruments: An Essay on 'Science,' 'Religion' and Applied Metaphysics," *Studies in History and Philosophy of Science* 38, no. 2 (2007): 374. See also chapter 2 in Thomas L. Hankins and Robert J. Silverman, *Instruments and the Imagination* (Princeton: Princeton University Press, 1995).

58. Translations (with modifications) taken from Auguste Villiers de L'Isle-Adam, *Tomorrow's Eve*, trans. Robert Martin Adams (Urbana: University of Illinois Press, 1982), 8.

59. Ibid., 8–13.

60. Ibid., 19–20.

61. Friedrich Kittler analyses the same passage of *L'Ève future* as an early and "concise" formulation of the epistemological transition triggered by the advent of media technology, which impelled us to rethink the nature of literature and the sacred. He writes, "The ability to record sense data technologically shifted the entire discourse network circa 1900. For the first time in history, writing ceased to be synonymous with the serial storage of data. The technological recording of the real entered into competition with the symbolic registration of the Symbolic." Friedrich A. Kittler, *Discourse Networks 1800/1900*, trans. Michael Metteer and Chris Cullens (Stanford: Stanford University Press, 1990), 229–30.

62. Villiers de L'Isle-Adam, *Tomorrow's Eve*, 36.

63. Ibid., 40.

64. Ibid., 46.

65. Ibid., 22, 54.

66. Ibid., 187, 213.

67. Ibid., 86.

68. Ibid., 110.

69. Ibid., 59–61.

70. Ibid., 130–31, 39–47, 50, 215.

71. Ibid., 131.

72. Ibid., 204.

73. Justinus Kerner, *The Seeress of Prevorst: Being Revelations Concerning the Inner-Life of Man, and the Inter-diffusion of a World of Spirits in the One We Inhabit*, trans. Catherine Crowe (London: J. C. Moore, 1845), 96–97.

74. Villiers de L'Isle-Adam, *Tomorrow's Eve*, 209–10. The actual Edison undertook to communicate with spirits by building machines "so delicate as to be affected, or moved, or manipulated [. . .] by our personality as it survives in the next life." Cited in Linda Simon, *Dark Light: Electricity and Anxiety from the Telegraph to the X-Ray* (Orlando: Harcourt, 2004), 187.

75. As the fictional Edison reflects,

> The moral quality that I recognized in Mrs. Anderson before her illness, and that which I discover at the depths of her hypnotic slumber [*la profondeur magnétique*], seem to me absolutely distinct. She used to be a simple woman, perfectly honorable, even intelligent, but, after all, of very limited views—and so I knew her. But in the depths of her slumber another person is

revealed to me, completely different, many-sided and mysterious! So far as I can tell, the enormous knowledge, the strange eloquence, and the penetrating insight of this sleeper named Sowana—who is, physically, the same person—are logically inexplicable. Isn't this duality a stupefying phenomenon? And yet this same duality, though lesser in degree and not quite so striking, is a regularly observed, a recognized phenomenon in almost all the subjects treated by trained investigators. Sowana is exceptional only as an abnormally perfect instance of a very common event—and this abnormality is due simply to her particular variety of neurosis. (Villiers de L'Isle-Adam, *Tomorrow's Eve*, 210–11)

76. Ibid., 211.

77. Villiers makes this clear by calling Mister Edward Anderson's overvalued lover "Miss *Eve*lyn Habal."

78. The incorporation of Sowana could be said to be the symptom of a "crypt," which Derrida defines as "a foreigner in the Self, and especially of the heterocryptic ghost that *returns* from the Unconscious *of* the other"; or, "the crypt from which the ghost comes back belongs to someone else." Nicolas Abraham, Maria Torok, and Jacques Derrida, *The Wolf Man's Magic Word: A Cryptonymy* (Minneapolis: University of Minnesota Press, 1986), xxxi, 119.

79. Villiers de L'Isle-Adam, *Tomorrow's Eve*, 211.

80. Ibid., 216.

81. Ibid., 219.

82. Ibid., 213.

83. For a detailed analysis of hysteria in the novel, see Asti Hustvedt, "The Pathology of Eve: Villiers de l'Isle-Adam and Fin de Siècle Medical Discourse," in *Jeering Dreamers: Villiers de l'Isle-Adam's L'Eve Future at Our Fin de Siècle: A Collection of Essays*, ed. John Anzalone (Amsterdam: Rodopi, 1996).

84. Villiers de L'Isle-Adam, *Tomorrow's Eve*, 213–14.

85. Ibid.

86. Georges Canguilhem, *La connaissance de la vie*, Poche (Paris: Vrin, 2009), 134–35.

87. Ibid., 149. Today's programmers of self-learning AI might argue otherwise.

88. Sigmund Freud and Joseph Breuer, *Studies on Hysteria*, ed. and trans. James Strachey, Standard Edition of the Complete Psychological Works of Sigmund Freud, vol. 2 (London: Hogarth, 1955), 193–94.

89. Ibid., 203–4, 21.

90. Akira Mizuta Lippit, *Electric Animal: Toward a Rhetoric of Wildlife* (Minneapolis: University of Minnesota Press, 2000), 101–21.

91. Lippit writes,

> The unconscious, which forms, according to psychoanalysis, the subject's core, is in fact external to the subject and can only return to claim the subject as a supplement. There can thus be no self-sufficiency, autonomy, or closure in the world of consciousness without its completion by something foreign to it. Breuer has, in essence, discovered the radical alterity of the unconscious by realizing that its figuration, its supplement, determines the place of its incarnation. The supplement, trope, or figure of the unconscious is in fact the unconscious. The origin of the unconscious is not in the subject but outside it. In the first instance, this excess is marked by the electrical system and its magnetic, prelinguistic efficacy. (Ibid., 113–14)

Conclusion

1. Gaston Bachelard, *La formation de l'esprit scientifique*, Bibliothèque des textes philosophiques (Paris: Vrin, 2004).

2. Gaston Bachelard, "La vocation scientifique et l'âme humaine," in *L'homme devant la science (Rencontre internationale de Genève 1952)* (Neuchâtel: Les Éditions de la Baconnière, 1952), 15, 25–26.

3. Gaston Bachelard, *Le nouvel esprit scientifique* (Paris: F. Alcan, 1934), 42, 146–47.

4. Hans-Jörg Rheinberger, *On Historicizing Epistemology: An Essay* (Stanford: Stanford University Press, 2010).

5. Although not as influential, a philosophy of history inspired by electromagnetism had already been elaborated by Henry Adams. According to Sam Halliday, "The most famous episode in Adam's *The Education of Henry Adams* (1907) sees Adams encountering a dynamo, and being left with his own 'historical neck broken.' Coming from a professional historian, this break is doubly significant; signaling both a turning point in Adams's life story, and a *methodological* shift, away from conventional historicism, toward a science-based alternative whose guiding concepts, it turns out, are directly inspired by the dynamo itself. The dynamo is thus revealed (in James Carey's words, about the telegraph) as 'a thing to think with, an agency for the alteration of ideas'" (*Science and Technology in the Age of Hawthorne, Melville, Twain, and James: Thinking and Writing Electricity* [New York: Palgrave Macmillan, 2007], 45).

6. Charles Alunni, "Relativités et puissances spectrales chez Gaston Bachelard," *Revue de synthèse* 120, no. 1 (1999). See also Vincent Bontems, *Bachelard*, Figures du savoir (Paris: Les Belles lettres, 2010), 22–24, 124–26.

7. "Le monde réel et le *déterminisme dynamique* qu'il implique demandent d'autres *intuitions*, des *intuitions dynamiques* pour lesquelles il faudrait un nouveau vocabulaire philosophique. Si le mot induction n'avait déjà tant de sens, nous proposerions de l'appliquer à ces intuitions dynamiques." Gaston Bachelard, *L'activité rationaliste de la physique contemporaine* (Paris: Presses universitaires de France, 1951), 214. My translation.

8. In the original, "Il n'y a donc pas de transition entre le système de Newton et le système d'Einstein. On ne va pas du premier au second en amassant des connaissances, en redoublant de soins dans les mesures, en rectifiant légèrement des principes. Il faut au contraire un effort de nouveauté totale. On suit donc une induction transcendante et non pas une induction amplifiante en allant de la pensée classique à la pensée relativiste" (Bachelard, *Le nouvel esprit*, 42). Translation with modifications from Gaston Bachelard, *The New Scientific Spirit*, trans. Arthur Goldhammer (Boston: Beacon, 1984), 44.

9. Bachelard, *Le nouvel esprit*, 125.

10. Bachelard was familiar with Gratry's *Logique*. See the last page of Gaston Bachelard, *Le rationalisme appliqué*, 3rd ed. (Paris: Les Presses Universitaires de France, 1966).

11. Auguste Joseph Alphonse Gratry, *Logique*, vol. 2 (Paris: Lecoffre, 1855), 74, 76.

12. Ibid., 171–72.

13. "But later investigations [. . .] of the laws governing these phenomena, induce me to think that [. . .]" (Michael Faraday, "On the Induction of Electric Currents [. . .]," in *Experimental Researches in Electricity* [1839–55], 16). "Thus the reasons which induce me to suppose a particular state in wire (60.) have disappeared [. . .]" (Faraday, "Terrestrial Magneto-electric Induction [. . .]," in *Experimental Researches in Electricity*, [London: R. and J. E. Taylor, 1839–55], 69). As discussed above, similar wide-range invocations of *to induce* appeared in Cavallo's early treatise on electrostatics.

14. Bachelard himself lends support to this claim in the preface he wrote for Balzac's *Séraphîta*. In this novel, Balzac explores human and divine interactions in terms of images inspired by Swedenborgian correspondences and, as we saw, electromagnetism. Although he does not specifically refer to the electromagnetic ones, Bachelard describes the power of these images as instances of "dynamic induction." Gaston Bachelard, *Le droit de rêver* (Paris: Presses Universitaires de France, 1970), 128.

15. In the original, "Et les critiques portent souvent plus d'attention au mot qu'à la phrase—à la locution plus qu'à la page. Ils pratiquent un jugement essentiellement atomique et statique. Rares sont les critiques qui essaient un nouveau style en se soumettant à son *induction*. J'imagine, en effet, que de l'auteur au lecteur devrait jouer une *induction verbale* qui a bien des caractères de l'induction électromagnétique entre deux circuits. Un livre serait alors un appareil d'induction psychique qui devrait provoquer chez le lecteur des tentations d'expressions

originales" (ibid., 181). This citation is from an essay on Jean Paulhan initially published 1942–43. Translation (with modifications) from Gaston Bachelard, *The Right to Dream*, trans. J. A. Underwood (Dallas: Dallas Institute Publications, 1988).

16. Gaston Bachelard, *La psychanalyse du feu*, Folio essais ed. (Paris: Gallimard, 1987), 60.

17. Gaston Bachelard, *La terre et les rêveries du repos* (Paris: J. Corti, 1948), 293.

18. Gaston Bachelard, *L'eau et les rêves: Essai sur l'imagination de la matière* (Paris: J. Corti, 1956), 251–52.

19. Bachelard, *La psychanalyse*, 30.

20. Ibid., 109.

21. Ibid., 185–90.

22. "détacher tous les suffixes de la beauté, s'évertuer à trouver, derrière les images qui se montrent, les images qui se cachent, aller à la racine même de la force imaginante" (Bachelard, *L'eau et les rêves*, 3).

23. Julien Gracq, "André Breton, quelques aspects de l'écrivain," in *Œuvres complètes* (Paris: Gallimard, 1989), 429–30. André Breton, *Œuvres complètes*, ed. Marguerite Bonnet, Bibliothèque de la pléiade (Paris: Gallimard, 1988).

24. Gracq, "André Breton," 427–30.

25. Ibid., 423–26.

26. In the original, "Une telle image semble douée d'un pouvoir révélateur sur cette région de nous-mêmes où s'enracinent les attirances et les répugnances involontaires. Il semble que nous soyons en présence d'une pure image motrice, fort ancienne, née d'un état de besoin, mais depuis sa naissance obstinément en quête d'un correspondant matériel concret, qu'elle se sent d'avance en droit d'exiger du monde extérieur et en assurance de découvrir malgré ses mécomptes, soit qu'elle le cherche d'abord avec Mesmer du côté du 'magnétisme animal' ou avec Goethe sous le nom d''affinités électives' dans le domaine de la chimie" (ibid., 424–25). Many thanks to Garret T. Murphy for helping me with the translations.

27. In the original, "L' 'homme' du XVIIe siècle peut bien nous être connu, [. . .] il reste que jamais le *courant* n'y passe pour laisser apparaître tout son cortège de phénomène induits—jamais rien ne se laisse entrevoir de cette danse sacrée qui s'improvise du contact de deux êtres et dont leur comportement ultérieur ne fera que développer jusqu'à la lassitude des variations—jamais à travers cette monade déifié, le *tout* de l'homme, qui est d'être un moi à jamais pris dans le tourbillon magnétique d'un système de 'moi' [. . .] n'arrive à prendre le moindre caractère de réalité" (ibid., 425).

28. In the original,

> Par apport à cette psychologie purement mécanique, Dostoïevski incontestablement se situerait comme l'autre pôle: à peine ouvre-t-on un livre comme *Les Possédés* [. . .] que nous nous trouvons jetés au sein d'un monde irradié, traversé, par l'entremise d'un moi exceptionnellement bon conducteur, d'influx et de lignes de force. La figure centrale de Stavroguine s'y montre douée non d'une structure mentale particulièrement différenciée, mais du caractère à la fois impénétrable et fascinant des manifestations électriques: pour la première fois sans doute avec cette netteté [. . .] nous est présenté un être dont la réalité, pourtant plus grande que nature, se réduit expressément à son seul *sillage*; aux phénomènes de turbulence, de décomposition et de recomposition accélérée qu'il engendre sur son parcours. (Ibid., 425–26)

29. Iwan Rhys Morus, *When Physics Became King* (Chicago: University of Chicago Press, 2005), 275–77.

30. Gustave Le Bon, *Psychologie des foules* (Paris: Presses Universitaires de France, 2008), 13–14, 76.

31. Gustave Le Bon, *L'évolution de la matière* (Paris: Flammarion, 1905). See also Henderson, *Duchamp in Context: Science and Technology in the Large Glass and Related Works*, 7.

32. Gracq, "André Breton," 326.

33. Le Bon, *Psychologie des foules*, 59–63.

34. Ibid., 60.

35. See the *bouton d'appel* entry in Julien Lefévre, *Dictionnaire d'électricité comprenant*

les applications aux sciences (Paris: J. B. Baillière et fils, 1895), 704, 832.

36. In the original, "En desserrant de son mieux les règles mécaniques d'assemblage de mots, en les libérant des attractions banales de la logique et de l'habitude, en les laissant 'tomber' dans un vide intérieur [. . .], il observera et suivra aveuglément entre eux de secrètes attractions magnétiques, il laissera 'les mots faire l'amour' et un *monde* insolite finalement se recomposer à travers eux en liberté. [. . .] il assistera du dehors, en spectateur, à l'élaboration spontanée de cette magie continuelle, se bornant à signer, par un acte dont la gratuité comportera toujours une part d'humour, les cristallisations les plus réussies" (Gracq, "André Breton," 477).

37. In the original,

> L'affleurement d'une pensée à la conscience participe toujours plus ou moins du caractère émouvant de la naissance—intellectuelle, certes, dans son essence, elle est mêlée indéfinissablement à une frange colorée d'affectivité, éminemment vibratile, capable de propager des ondes dans les zones voisines, elle est un élan moteur et potentiel magnétique capable d'infinies et brusques variations. Or, il est clair que cet état naissant d'une pensée, vague, protéiforme, mais douée d'une charge affective considérable (la pensée faite vibration), est entre tous celui qui paraît le plus apte à la communication. Certains regards ou certains gestes d'un acteur inspiré, certains *mouvements* d'un grand poète, avant toute élucidation d'un contenu, éveillent notre sensibilité—par opposition à cette menue monnaie mentale, si aisément la proie des faussaires, que remue la "mise en circulation des idées"—à des états *contagieux* de la pensée. (Ibid., 475–76)

38. In the original,

> Attachée à ce moment de mystère où l'esprit se fait pur épanouissement, on sait de reste que la poésie ne se propose jamais rien tant, par la rupture de toutes associations habituelles au moyen de l'image, que de provoquer artificiellement cet état naissant en essayant de nous faire voir chaque objet dans une lumière de création du monde et comme pour la première fois. Dans l'état de grâce poétique, que ce soit ou non une illusion, le courant semble passer de conscience en conscience sans obstacle, une entrée en résonnance spontanée se produit qui donne plus encore que l'impression de la communication celle de la "co-naissance" pour reprendre l'expression de Claudel. (Ibid., 477)

39. Paul Claudel, *Œuvre poétique*, Bibliothèque de la pléiade (Paris: Gallimard, 1967). Gilles Deleuze inscribes Claudel's notion of co-naissance in the Aristotelian epistemological tradition, which consists in grounding the formation of knowledge on an intimate complicity between mind and matter, thing and concept, or man and nature (or man and the world). Following the "Kantian rupture," this complicity ceased to supply the foundation of modern epistemologies and was only to be recovered in the twentieth century with notions such as Husserl's "being in the world." From Aristotle to Claudel, what makes knowledge possible is the universality of movement. Gilles Deleuze, "Cinéma / Pensée" (lecture, Université Paris 8: La voix de Gilles Deleuze en ligne, 6–2019 1984), http://www2.univ-paris8.fr/deleuze/article.php3?id_article=371.

Bibliography

Abraham, Nicolas, Maria Torok, and Jacques Derrida. *The Wolf Man's Magic Word: A Cryptonymy*. Minneapolis: University of Minnesota Press, 1986.

Adler, Jeremy. "The Aesthetics of Magnetism: Science, Philosophy, and Poetry in the Dialogue Between Goethe and Schelling." In *The Third Culture: Literature and Science*, edited by Elinor S. Shaffer, 66–102. European Cultures. Berlin: Walter de Gruyter, 1998.

Alunni, Charles. "Relativités et puissances spectrales chez Gaston Bachelard." *Revue de synthèse* 120, no. 1 (1999): 73–110.

Ambrière, Madeleine. *Balzac et la recherche de l'absolu*. Paris: Hachette, 1968.

Ampère, André-Marie. *Théorie des phénomènes électro-dynamiques, uniquement déduite de l'expérience*. Paris: Méquignon-Marvi, 1826.

Andrä, Wilfried, and Hannes Nowak. *Magnetism in Medicine: A Handbook*. 2nd ed. Weinheim, Germany: Wiley-VCH, 2007.

Andrade Martins, Roberto de. "Ørsted, Ritter, and Magnetochemistry." In *Hans Christian Ørsted and the Romantic Legacy in Science Ideas, Disciplines, Practices*, edited by Robert Michael Brain, R. S. Cohen, and Ole Knudsen, 339–85. Dordrecht: Springer, 2007.

Anonymous. "Notices Respecting New Books." *Philosophical Magazine* 40 (1812): 145–51, 297–308, 434–44.

Aristotle. "On the Soul." Internet Classics Archive. Accessed June 2019. http://classics.mit.edu/Aristotle/soul.1.i.html.

Bachelard, Gaston. *The Formation of the Scientific Mind: A Contribution to a Psychoanalysis of Objective Knowledge*. Translated by Mary McAllester Jones. Manchester, England: Clinamen Press, 2002.

———. *L'activité rationaliste de la physique contemporaine*. Paris: Presses universitaires de France, 1951.

———. *La formation de l'esprit scientifique*. Bibliothèque des textes philosophiques. 1938. Reprint, Paris: Vrin, 2004.

———. *La psychanalyse du feu*. Folio essais ed. Paris: Gallimard, 1987.

———. *La terre et les rêveries du repos*. Paris: J. Corti, 1948.

———. "La vocation scientifique et l'âme humaine." In *L'homme devant la science (Rencontre internationale de Genève 1952)*, 11–30. Neuchâtel, Switzerland: Éditions de la Baconnière, 1952.

———. *L'eau et les rêves: Essai sur l'imagination de la matière*. 1942. Reprint, Paris: J. Corti, 1956.

———. *Le droit de rêver*. Paris: Presses universitaires de France, 1970.

———. *Le nouvel esprit scientifique*. Paris: F. Alcan, 1934.

———. *Le rationalisme appliqué*. 3rd ed. Paris: Presses universitaires de France, 1966.

———. *The New Scientific Spirit*. Translated by Arthur Goldhammer. Boston: Beacon Press, 1984.

———. *The Right to Dream*. Translated by J. A. Underwood. Dallas: Dallas Institute Publications, 1988.

Bacon, Francis. *The New Organon*. Edited by Lisa Jardine and Michael Silverthorne. Cambridge: Cambridge University Press, 2000.

Baldwin, Martha. "Athanasius Kircher and the Magnetic Philosophy." Ph.D. diss., University of Chicago, 1987.

Balibar, Françoise. *Einstein 1905: De l'éther aux quanta*. Paris: Presses universitaires de France, 1992.

Balzac, Honoré de. "Avant-propos." In *La comédie humaine*. Bibliothèque de la pléiade, vol. 1, 7–20. Paris: Gallimard, 1976.

———. *Correspondance*. Edited by Roger Pierrot. Vol. 2. Paris: Éditions Garnier Frères, 1960.

———. *Histoire intellectuelle de Louis Lambert, par M. de Balzac, fragment extrait des "Romans et contes philosophiques."* Paris: Charles Gosselin, 1833.

———. *La recherche de l'absolu*. Folio Classique. Paris: Gallimard, 2005.

———. *Le père Goriot*. Paris: Livre de Poche, 1983.

———. *Lettres à l'étrangère*. Paris: Calmann-Lévy, 1906.

———. *Louis Lambert*. In *La comédie humaine*. Bibliothèque de la pléiade, vol. 10. Paris: Gallimard, 1935.

———. *Nouveaux contes philosophiques, par M. de Balzac. Maître Cornélius; Madame Firmiani; L'Auberge rouge; Louis Lambert*. Paris: Charles Gosselin, 1832.

———. *Séraphita*. In *La comédie humaine*. Bibliothèque de la pléiade, vol. 11. Paris: Gallimard, 1976.

———. *Théorie de la démarche*. In *La comédie humaine*. Bibliothèque de la pléiade, vol. 12. Paris: Gallimard, 1981.

———. *Ursule Mirouët*. In *La comédie humaine*. Bibliothèque de la pléiade, vol. 3. Paris: Gallimard, 1935.

Barcelona, Antonio, ed. *Metaphor and Metonymy at the Crossroads: A Cognitive Perspective*. Berlin: de Gruyter, 2012.

Bierwiaczonek, Boguslaw. *Metonymy in Language, Thought and Brain*. Sheffield, England: Equinox, 2013.

Blix, Göran. "The Occult Roots of Realism: Balzac, Mesmer, and Second Sight." *Studies in Eighteenth-Century Culture* 36 (2007): 261–80.

Blondel, Christine. *A.-M. Ampère et la création de l'électrodynamique, 1820–1827*. Paris: Bibliothèque nationale, 1982.

Blondel, Christine, and Bertrand Wolff. "@. Ampère et l'histoire de l'électricité." Accessed June 2019. http://www.ampere.cnrs.fr/.

Bontems, Vincent. *Bachelard*. Figures du savoir. Paris: Les Belles Lettres, 2010.

Breton, André. *Œuvres complètes*. Edited by Marguerite Bonnet. Bibliothèque de la pléiade, vol. 3. Paris: Gallimard, 1988.

Burke, Kenneth. "Four Master Tropes." *Kenyon Review* 3, no. 4 (1941): 421–38.

Butler, Marilyn. "'Frankenstein' and Radical Science." In *Frankenstein*, edited by J. Paul Hunter, 302–13. Norton Critical Edition. New York: Norton, 1996.

Bynum, William F. "The Great Chain of Being After Forty Years: An Appraisal." *History of Science* 13 (1975): 1–28.

Caneva, Kenneth L. "'Discovery' as a Site for the Collective Construction of Scientific Knowledge." *Historical Studies in the Physical and Biological Sciences* 35, no. 2 (2005): 175–291.

———. "Physics and *Naturphilosophie*: A Reconnaissance." *History of Science* 35, no. 107 (1997): 35–106.

Canguilhem, Georges. *La connaissance de la vie*. Poche. 1965. Reprint, Paris: Vrin, 2009.

Carnot, Sadi. *Réflexions sur la puissance motrice du feu et sur les machines propres à développer cette puissance*. Paris: Bachelier, 1824.

Cavallo, Tiberius. *A Complete Treatise of Electricity in Theory and Practice; With Original Experiments*. London: Edward and Charles Dilly, 1777.

Chaucer, Geoffrey. "Canterbury Tales." Edited by Gerard NeCastro. Medievalit. Accessed June 2019. https://medievalit.com/home/echaucer/original-texts/the-canons-yeomans-tale-original-text/.

Chertok, Léon, and Raymond de Saussure. *The Therapeutic Revolution, from Mesmer to Freud*. New York: Brunner/Mazel, 1979.

Clarke, Bruce. *Energy Forms: Allegory and Science in the Era of Classical Thermodynamics*. Ann Arbor: University of Michigan Press, 2001.

Claudel, Paul. *Œuvre poétique*. Bibliothèque de la pléiade. Paris: Gallimard, 1967.

Cobb, Aaron D. "History and Scientific Practice in the Construction of an Adequate Philosophy of Science: Revisiting a Whewell/Mill Debate." *Studies in History and Philosophy of Science* 42, no. 1 (2011): 85–93.

Conceptual Metonymy: Methodological, Theoretical, and Descriptive Issues.

Edited by Olga Blanco-Carrión, Antonio Barcelona, and Rossella Pannain. Amsterdam: John Benjamins, 2018.

Craig, Cairns. *Associationism and the Literary Imagination: From the Phantasmal Chaos*. Edinburgh: Edinburgh University Press, 2007.

Craighill, Stephanie. "The Influence of Duality and Poe's Notion of the 'Bi-Part Soul' on the Genesis of Detective Fiction in the Nineteenth-Century." Ph.D. diss., Edinburgh Napier University, 2010.

Darnton, Robert. *Mesmerism and the End of the Enlightenment in France*. Cambridge: Harvard University Press, 1968.

Davy, Humphry. *Elements of Chemical Philosophy*. Vol. 1. Philadelphia: Bradford and Inskeep, 1812.

De Bruno, M. *Recherches sur la direction du fluide magnétique*. Amsterdam: Chez Gueffier, 1785.

De Dobay Rifelj, Carol. "Minds, Computers and Hadaly." In *Jeering Dreamers: Villiers de l'Isle-Adam's L'Eve Future at Our Fin de Siècle: A Collection of Essays*, edited by John Anzalone, 127–39. Amsterdam: Rodopi, 1996.

De Fontenelle, Jean-Sébastien-Eugène Julia, and François Malepeyre. *Nouveau manuel complet de physique amusante ou nouvelles récréations physiques*. Paris: Roret, 1860.

Delbourgo, James. *A Most Amazing Scene of Wonders: Electricity and Enlightenment in Early America*. Cambridge: Harvard University Press, 2006.

Deleuze, Gilles. "Cinéma / Pensée." Université Paris 8: La voix de Gilles Deleuze en ligne. Accessed June 2019. http://www2.univ-paris8.fr/deleuze/article.php3?id_article=371.

Deleuze, Joseph Philippe François. *Practical Instruction in Animal Magnetism, Part 1*. Translated by Thomas C. Hartshorn. 2nd ed. Providence, R.I.: B. Cranston, 1837.

Derrida, Jacques. *De la grammatologie*. Paris: Éditions de Minuit, 1997.

Descartes, René. *Principles of Philosophy*. Translated by Valentine Rodger Miller and Reese P. Miller. 1644. Reprint, Dordrecht: D. Reidel, 1983.

Doane, Mary Ann. "Technophilia: Technology, Representation, and the Feminine." In *The Gendered Cyborg: A Reader*, edited by Gill Kirkup, Linda Janes, Kathryn Woodward, and Fiona Hovenden, 110–21. London: Routledge, 2000.

Du Potet, Jules Denis. "Électro-Magnétisme." *Journal du magnétisme* 12 (1853): 201–21.

Dusek, Val. *The Holistic Inspirations of Physics: The Underground History of Electromagnetic Theory*. New Brunswick: Rutgers University Press, 1999.

Eco, Umberto, Thomas A. Sebeok, Nancy Harrowitz, Jean Umiker-Sebeok, and Carlo Ginzburg. *The Sign of Three: Dupin, Holmes, Peirce*. Edited by Umberto Eco and

Thomas A. Sebeok. Advances in Semiotics. Bloomington: Indiana University Press, 1983.

Edelman, Nicole. *Voyantes, guérisseuses et visionnaires en France: 1785–1914*. Paris: A. Michel, 1995.

Einstein, Albert. "Excerpt from Essay by Einstein on 'Happiest Thought' in His Life." *New York Times*, March 28, 1972.

———. "Field Theories, Old and New." *New York Times*, February 3, 1929.

———. "On the Electrodynamics of Moving Bodies." Translated by Anna Beck. In *The Swiss Years: Writings, 1900–1909*, 140–71. The Collected Papers of Albert Einstein (English Translation Supplement). Princeton: Princeton University Press, 1989.

Faivre, Antoine. "Borrowings and Misreading: Edgar Allan Poe's 'Mesmeric' Tales and the Strange Case of Their Reception." *Aries* 7 (2007): 21–62.

Fara, Patricia. *Fatal Attraction: Magnetic Mysteries of the Enlightenment*. New York: MJF Books / Fine Communications, 2006.

———. *Sympathetic Attractions: Magnetic Practices, Beliefs, and Symbolism in Eighteenth-Century England*. Princeton: Princeton University Press, 1996.

Faraday, Michael. "On Lines of Magnetic Force [. . .]." In *Experimental Researches in Electricity*, 328–70. London: R. and J. E. Taylor, 1839–1855.

———. "On Some Points of Magnetic Philosophy." In *Experimental Researches in Electricity*, 566–74. London: R. and J. E. Taylor, 1839–1855.

———. "On Table-Turning." In *Experimental Researches in Chemistry and Physics*, 382–91. London: R. Taylor and W. Francis, 1859.

———. "On the Induction of Electric Currents [. . .]." In *Experimental Researches in Electricity*, 1–41. London: R. and J. E. Taylor, 1839–1855.

———. "Terrestrial Magneto-electric Induction [. . .]." In *Experimental Researches in Electricity*, 42–75. London: R. and J. E. Taylor, 1839–1855.

Frank, Adam. "Valdemar's Tongue, Poe's Telegraphy." *ELH* 72, no. 3 (Fall 2005): 635–62.

Franklin, Benjamin. *Memoirs of Benjamin Franklin*. Edited by William Temple Franklin, William Duane, George B. Ellis, and Henry Stevens. 2 vols. Philadelphia: M'Carty & Davis, 1834.

Freud, Sigmund, and Joseph Breuer. *Studies on Hysteria*. Edited and Translated by James Strachey. The Standard Edition of the Complete Psychological Works of Sigmund Freud. Vol. 2. London: Hogarth Press, 1955.

Friedman, Michael. "Kant—Naturphilosophie—Electromagnetism." In *Hans Christian Ørsted and the Romantic Legacy in Science Ideas, Disciplines, Practices*, edited by Robert Michael Brain, R. S. Cohen, and Ole Knudsen, 135–58. Dordrecht: Springer, 2007.

Fumaroli, Marc. "Histoire de la littérature et histoire de l'électricité: Actes du colloque de l'Association pour l'histoire de l'électricité en France, Paris, 11–13 octobre 1983." In *L'Électricité dans l'histoire, Problèmes et méthodes*. Paris: Presses universitaires de France, 1985.

Galvani, Luigi. "De Viribus Electricitatis." In *Literature and Science in the Nineteenth Century: An Anthology*, edited by Laura Otis, 135–40. Oxford: Oxford University Press, 2002.

Gauld, Alan. *A History of Hypnotism*. Cambridge: Cambridge University Press, 1992.

Gautier, Théophile. *Portraits contemporains*. In *Oeuvres complètes*, vol. 9. Geneva, Switzerland: Slatkine Reprints, 1978.

Gelfert, Axel. "Observation, Inference, and Imagination: Elements of Edgar Allan Poe's Philosophy of Science." *Science and Education* 23, no. 3 (2014): 589–607.

Genette, Gérard. "La rhétorique restreinte." *Communications*, no. 16 (1970): 158–71.

Gilbert, William. *On the Magnet*. New York: Basic Books, 1958.

Gilmore, Paul. *Aesthetic Materialism: Electricity and American Romanticism*. Stanford: Stanford University Press, 2009.

Gold, Barri J. *ThermoPoetics: Energy in Victorian Literature and Science*. Cambridge: MIT Press, 2010.

Gooday, Graeme. *Domesticating Electricity: Technology, Uncertainty, and Gender, 1880–1914*. Science and Culture in the Nineteenth Century. London: Pickering & Chatto, 2008.

Gooding, David C. "From Phenomenology to Field Theory: Faraday's Visual Reasoning." *Perspectives on Science* 14, no. 1 (2006): 40–65.

Gordon, Rae Beth. "Poe: Optics, Hysteria, and Aesthetic Theory." *Cercles* 1 (2000): 49–60.

Goulet, Andrea. *Legacies of the Rue Morgue: Science, Space, and Crime Fiction in France*. Critical Authors and Issues. Philadelphia: University of Pennsylvania Press, 2016.

Gracq, Julien. "André Breton, quelques aspects de l'écrivain." In *Œuvres complètes*, 397–515. Paris: Gallimard, 1989.

Gratry, Auguste Joseph Alphonse. *Logique*. Vol. 2. Paris: Lecoffre, 1855.

Griffiths, Devin. *The Age of Analogy: Science and Literature Between the Darwins*. Baltimore: Johns Hopkins University Press, 2016.

Grimstad, Paul. "C. Auguste Dupin and Charles S. Peirce: An Abductive Affinity." *Edgar Allan Poe Review* 6, no. 2 (2005): 22–30.

Halley, Edmond. "An Account of the Late Surprizing Appearance of the Lights Seen in the Air, on the Sixth of March Last; with an

Attempt to Explain the Principal Phaenomena Thereof; as It Was Laid Before the Royal Society." *Philosophical Transactions* 29, no. 347 (1716): 406–28.

———. "A Discourse Concerning Gravity, and Its Properties, Wherein the Descent of Heavy Bodies, and the Motion of Projects Is Briefly, but Fully Handled: Together with the Solution of a Problem of Great Use in Gunnery." *Philosophical Transactions*, no. 16 (1686): 3–21.

Halliday, Sam. *Science and Technology in the Age of Hawthorne, Melville, Twain, and James: Thinking and Writing Electricity*. New York: Palgrave Macmillan, 2007.

Hammoud, Saïd. *Mesmérisme et romantisme allemand: 1766–1829*. Sciences et société. Paris: L'Harmattan, 1994.

Hanegraaff, Wouter J. "A Woman Alone: The Beatification of Friederike Hauffe *née* Wanner, 1801–1829." In *Women and Miracle Stories*, edited by Anne-Marie Korte, 211–47. Leiden, Netherlands: Brill, 2000.

Hankins, Thomas L., and Robert J. Silverman. *Instruments and the Imagination*. Princeton: Princeton University Press, 1995.

Hartley, David. *Observations on Man, His Frame, His Duty, and His Expectations*. Vol. 1. London: S. Richardson, 1749.

Hegel, Georg Wilhelm Friedrich. *Le magnétisme animal: Naissance de l'hypnose*. Edited by François Roustang. Quadrige Grands Textes. Paris: Presses universitaires de France, 2005.

Heilbron, John L. *Electricity in the 17th and 18th Centuries: A Study of Early Modern Physics*. Berkeley: University of California Press, 1979.

Henderson, Linda Dalrymple. *Duchamp in Context: Science and Technology in the Large Glass and Related Works*. Princeton: Princeton University Press, 1998.

Hopwood, Nick, Simon Schaffer, and Jim Secord. "Seriality and Scientific Objects in the Nineteenth Century." *History of Science* 48, no. 3–4 (2010): 251–85.

Hughes, Thomas Parke. *Networks of Power: Electrification in Western Society, 1880–1930*. Baltimore: Johns Hopkins University Press, 1983.

Hume, David. "An Abstract of [. . .] a Treatise of Human Nature." In *A Treatise of Human Nature*, edited by David Fate Norton and Mary J. Norton, 403–17. Oxford Philosophical Texts. Oxford: Oxford University Press, 2000.

———. *A Treatise of Human Nature*. Edited by David Fate Norton and Mary J. Norton. Oxford Philosophical Texts. Oxford: Oxford University Press, 2000.

Hunt, Bruce J. *Pursuing Power and Light: Technology and Physics from James Watt to Albert Einstein*. Baltimore: Johns Hopkins University Press, 2010.

Hurh, Paul. "'The Creative and the Resolvent': The Origins of Poe's

Analytical Method." *Nineteenth-Century Literature* 66, no. 4 (2012): 466–93.

Hustvedt, Asti. "The Pathology of Eve: Villiers de l'Isle-Adam and Fin de Siècle Medical Discourse." In *Jeering Dreamers: Villiers de l'Isle-Adam's L'Eve Future at Our Fin de Siècle: A Collection of Essays*, edited by John Anzalone, 25–46. Amsterdam: Rodopi, 1996.

Jakobson, Roman. *Essais de linguistique générale*. Paris: Éditions de Minuit, 1963.

———. "Two Aspects of Language and Two Types of Aphasic Disturbances." In *Fundamentals of Language*, 53–82. London: J. F. and C. Rivington, 1785. Reprint, The Hague: Mouton, 1956.

Joule, James Prescott. *The Scientific Papers of James Prescott Joule*. 2 vols. London: Taylor and Francis, 1884.

Juhel, Pierre. *Histoire de la boussole: L'aventure de l'aiguille aimantée*. Beaux livres. Versailles: Éditions Quae, 2013.

Kahn, Douglas. *Earth Sound Earth Signal: Energies and Earth Magnitude in the Arts*. Berkeley: University of California Press, 2013.

Kang, Minsoo. *Sublime Dreams of Living Machines: The Automaton in the European Imagination*. Cambridge: Harvard University Press, 2011.

Kerner, Justinus. *The Seeress of Prevorst: Being Revelations Concerning the Inner-Life of Man, and the Inter-diffusion of a World of Spirits in the One We Inhabit*. Translated by Catherine Crowe. London: J. C. Moore, 1845.

Kittler, Friedrich A. *Discourse Networks 1800/1900*. Translated by Michael Metteer and Chris Cullens. Stanford: Stanford University Press, 1990.

Koyré, Alexandre. *Du monde clos à l'univers infini*. Paris: Gallimard, 2005.

Kuhn, Thomas. "Energy Conservation as an Example of Simultaneous Discovery." In *The Essential Tension: Selected Studies in Scientific Tradition and Change*, 66–104. Chicago: University of Chicago Press, 1959.

Lakoff, George, and Mark Johnson. *Metaphors We Live By*. Chicago: University of Chicago Press, 1980.

Lautréamont. *Les chants de Maldoror*. Paris: Flammarion, 1990.

Le Bon, Gustave. *L'évolution de la matière*. Paris: Flammarion, 1905.

———. *Psychologie des foules*. 1895. Reprint, Paris: Presses universitaires de France, 2008.

Lefévre, Julien. *Dictionnaire d'électricité comprenant les applications aux sciences*. Paris: J. B. Baillière et fils, 1895.

Lévy, Sydney. "Balzac et la machine-*peau de chagrin*." *Théorie, Littérature, Enseignement* 21 (2003): 133–52.

Lieberman, Jennifer L. *Power Lines: Electricity in American Life and Letters, 1882–1952*. Cambridge: MIT Press, 2017.

Linden, Stanton J. *Darke Hierogliphicks: Alchemy in English Literature from Chaucer to the Restoration*.

Lexington: University Press of Kentucky, 1996.

Lippit, Akira Mizuta. *Electric Animal: Toward a Rhetoric of Wildlife.* Minneapolis: University of Minnesota Press, 2000.

Littlemore, Jeannette. *Metonymy: Hidden Shortcuts in Language, Thought and Communication.* Cambridge: Cambridge University Press, 2015.

Lock, Charles. "Debts and Displacements: On Metaphor and Metonymy." *Acta Linguistica Hafniensia* 29, no. 1 (1997): 321–37.

Lodge, Oliver. *Life and Matter: A Criticism of Professor Haeckel's "Riddle of the Universe."* New York: G. P. Putnam's Sons, 1905.

Lovejoy, Arthur O. *The Great Chain of Being: A Study of the History of an Idea.* Cambridge: Harvard University Press, 1936.

Magee, Glenn Alexander. *Hegel and the Hermetic Tradition.* Ithaca: Cornell University Press, 2001.

Maricourt, Pierre de. *Letter of Petrus Peregrinus on the Magnet, A.D. 1269.* Translated by Brother Arnold. 1269. Reprint, New York: McGraw, 1904.

Markley, Robert. *Fallen Languages: Crises of Representation in Newtonian England, 1660–1740.* Ithaca: Cornell University Press, 1993.

Matzner, Sebastian. *Rethinking Metonymy: Literary Theory and Poetic Practice from Pindar to Jakobson.* Classics in Theory. Oxford: Oxford University Press, 2016.

Maxwell, James Clerk. "On Action at a Distance." In *The Scientific Papers of James Clerk Maxwell.* Vol. 2, 311–23. Cambridge: Cambridge University Press, 1890.

———. *A Treatise on Electricity and Magnetism.* Vol. 1. Oxford: Clarendon Press, 1873.

———. *A Treatise on Electricity and Magnetism.* Vol. 2. Oxford: Clarendon Press, 1873.

Meehan, Sean Ross. "Ecology and Imagination: Emerson, Thoreau, and the Nature of Metonymy." *Criticism* 55, no. 2 (2013): 299–329.

Méheust, Bertrand. *Somnambulisme et médiumnité: 1784–1930.* Collection Les empêcheurs de penser en rond. 2 vols. Le Plessis-Robinson: Institut Synthélabo pour le progrès de la connaissance, 1999.

Menke, Richard. *Telegraphic Realism: Victorian Fiction and Other Information Systems.* Stanford: Stanford University Press, 2008.

Mesmer, Franz Anton. *Mémoire de F. A. Mesmer [. . .] sur ses découvertes.* Paris: Fuchs, 1799.

Mesmer, Franz Anton, and Louis Caullet de Veaumorel. *Aphorismes de M. Mesmer [. . .].* Paris: M. Quinquet l'aîné, 1785.

Messac, Régis. *Le "Detective novel" et l'influence de la pensée scientifique.* Paris: Honoré Champion, 1929.

Mill, John Stuart. *A System of Logic, Ratiocinative and Inductive, Being a Connected View of the Principles of Evidence and the Methods of Scientific Investigation.* Vol. 1. London: John W. Parker, 1843.

Mills, Bruce. "Mesmerism." In *Edgar Allan Poe in Context*, edited by Kevin J. Hayes, 322–30. Cambridge: Cambridge University Press, 2013.

———. *Poe, Fuller, and the Mesmeric Arts: Transition States in the American Renaissance*. Columbia: University of Missouri Press, 2006.

Moldenhauer, Joseph J. "Poe's 'The Spectacles': A New Text from Manuscript Edited, with Textual Commentary and Notes." *Studies in the American Renaissance*, June 1977, 179–234.

Morus, Iwan Rhys. *When Physics Became King*. Chicago: University of Chicago Press, 2005.

Nancy, Jean-Luc. "Le Partage des Voix." In *Ion*, edited by Jean-François Pradeau and Édouard Mehl, 113–38. Paris: Ellipses, 2001.

Nancy, Jean-Luc, Mikkel Borch-Jacobsen, and Eric Michaud. *Hypnoses*. Paris: Editions Galilée, 1984.

Newton, Isaac. *The Mathematical Principles of Natural Philosophy*. Translated by Andrew Motte. Vol. 1. London: Benjamin Motte, 1729.

———. *Opticks: or, A Treatise of the Reflections, Refractions, Inflections and Colours of Light*. 4th ed. London: William Innys, 1730.

Niiniluoto, Ilkka. *Truth-Seeking by Abduction*. Cham, Switzerland: Springer, 2018.

Noiray, Jacques. *Jules Verne, Villiers de L'Isle-Adam*. Le Romancier et la machine. Vol. 2. Paris: Corti, 1982.

Nygaard, Loisa. "Winning the Game: Inductive Reasoning in Poe's 'Murders in the Rue Morgue.'" *Studies in Romanticism* 33, no. 2 (1994): 223–54.

Ogden, Emily. *Credulity: A Cultural History of US Mesmerism*. Edited by Kathryn Lofton and John Lardas Modern. Class 200: New Studies in Religion. Chicago: University of Chicago Press, 2018.

Orbaugh, Sharalyn. "Emotional Infectivity: Cyborg Affect and the Limits of the Human." In *Mechademia 3: Limits of the Human*, edited by Frenchy Lunning, 150–72. Minneapolis: University of Minnesota Press, 2008.

O'Reardon, John P., Murat Altinay, and Pilar Cristancho. "Transcranial Magnetic Stimulation: A New Treatment Option for Major Depression." *Psychiatric Times* 27, no. 9 (2010): 26–29.

Paavola, Sami. "Peircean Abduction: Instinct or Inference?" *Semiotica*, no. 153–1/4 (2005): 131–54.

Panther, Klaus-Uwe, and Günter Radden, eds. *Metonymy in Language and Thought*. Human Cognitive Processing. Amsterdam: Benjamins, 1999.

Panther, Klaus-Uwe, and Linda L. Thornburg. "Metonymy." In *The Oxford Handbook of Cognitive Linguistics*, edited by Dirk Geeraerts and H. Cuyckens, 236–63. Oxford: Oxford University Press, 2007.

———. "What Kind of Reasoning Mode Is Metonymy?" In *Conceptual Metonymy: Methodological, Theoretical, and Descriptive Issues*, edited by Olga Blanco-Carrión, Antonio Barcelona, and Rossella Pannain, 121–60. Amsterdam: John Benjamins, 2018.

Peacock, John. *The Look of Van Dyck: The Self-Portrait with a Sunflower and the Vision of the Painter*. Histories of Vision. Aldershot, England: Ashgate, 2006.

Pick, Daniel. *Svengali's Web: The Alien Enchanter in Modern Culture*. New Haven: Yale University Press, 2000.

Plato. "Ion." The Internet Classics Archive. Accessed June 2019. http://classics.mit.edu/Plato/ion.html.

Plato, M. Ficin, Johann Wolfgang von Goethe, and Jean-Luc Nancy. *Ion*. Edited by Jean-François Pradeau and Édouard Mehl. Paris: Ellipses, 2001.

Poe, Edgar Allan. *Eureka*. Edited by Stuart Levine and Susan F. Levine. Urbana: University of Illinois Press, 2004.

———. "The Facts in the Case of M. Valdemar." In *Collected Works of Edgar Allan Poe: Tales and Sketches, 1843–1849*, edited by Thomas Ollive Mabbott, 1228–44. Cambridge: Belknap Press of Harvard University Press, 1978. Accessed June 2019. http://www.eapoe.org/works/tales/vldmara.htm.

———. "Marginalia: Installment V: *Graham's Magazine*, March 1846." In *The Collected Writings of Edgar Allan Poe*, edited by Burton R. Pollin, 253–62. New York: Gordian Press, 1985. Accessed June 2019. http://www.eapoe.org/works/misc/mar0346.htm.

———. "Mesmeric Revelation." In *The Works of the Late Edgar Allan Poe*, edited by Rufus Wilmot Griswold, 110–20. New York: J. S. Redfield, Clinton Hall, 1844. Accessed June 2019. http://www.eapoe.org/works/TALES/mesmerd.htm.

———. "The Murders in the Rue Morgue." In *Collected Works of Edgar Allan Poe: Tales and Sketches, 1831–1842*, edited by Thomas Ollive Mabbott, 521–74. Cambridge: Belknap Press of Harvard University Press, 1978. Accessed June 2019. https://www.eapoe.org/works/mabbott/tom2t043.htm.

———. "The Narrative of Arthur Gordon Pym. Of Nantucket." In *The Collected Writings of Edgar Allan Poe*, edited by Burton R. Pollin, 53–363. New York: Gordian Press, 1994. Accessed June 2019. https://www.eapoe.org/works/pollin/brp1c000.htm.

———. "Poe to James R. Lowell, July 2, 1844." In *The Letters of Edgar Allan Poe*, edited by John Ward Ostrom, 255–59. New York: Gordian Press, 1966. Accessed June 2019. https://www.eapoe.org/works/ostlttrs/pl661c06.htm#pg0258.

———. "Poe to Philip P. Cooke, August 9, 1846." In *The Collected Letters of*

Edgar Allan Poe, edited by John Ward Ostrom, Burton R. Pollin, and Jeffrey A. Savoye, 594–98. New York: Gordian Press, 2008. Accessed June 2019. https://www.eapoe.org/works/ostlttrs/plo81co8.htm.

———. "Review of Human Magnetism." In *The Complete Works of Edgar Allan Poe*, edited by James Albert Harrison, 121–23. New York: Thomas Y. Crowell, 1845. Accessed June 2019. https://www.eapoe.org/works/harrison/jah12co8.htm.

———. *The Selected Writings of Edgar Allan Poe: Authoritative Texts, Backgrounds and Contexts, Criticism*. Edited by Gary Richard Thompson. Norton Critical Edition. Reprint, New York: W. W. Norton, 2004.

———. "The Spectacles." In *Collected Works of Edgar Allan Poe: Tales and Sketches, 1843–1849*, edited by Thomas Ollive Mabbott, 883–918. Cambridge: Belknap Press of Harvard University Press, 1978. Accessed June 2019. https://www.eapoe.org/works/mabbott/tom3too7.htm.

Pollin, Burton R. "'The Spectacles' of Poe—Sources and Significance." *American Literature* 37, no. 2 (1965): 185–90.

Porter, Dahlia. *Science, Form, and the Problem of Induction in British Romanticism*. Edited by James Chandler. Cambridge Studies in Romanticism. Cambridge: Cambridge University Press, 2018.

Pumfrey, Stephen. *Latitude and the Magnetic Earth*. New York: MJF Books / Fine Communications, 2006.

Ramalingam, Chitra. "Natural History in the Dark: Seriality and the Electric Discharge in Victorian Physics." *History of Science* 48, no. 3–4 (2010): 371–98.

Rausky, Franklin. *Mesmer: Ou, la révolution thérapeutique*. Paris: Payot, 1977.

Rheinberger, Hans-Jörg. *On Historicizing Epistemology: An Essay*. Stanford: Stanford University Press, 2010.

Rickels, Laurence A. *Aberrations of Mourning*. Minneapolis: University of Minnesota Press, 2011.

Rimbaud, Arthur. *Œuvres*. Edited by Suzanne Bernard and André Guyaux. Classiques Garnier. Paris: Garnier Frères, 1983.

Rimbaud, Arthur, Wallace Fowlie, and Seth Adam Whidden. *Rimbaud: Complete Works, Selected Letters: A Bilingual Edition*. Translated by Wallace Fowlie. Chicago: University of Chicago Press, 2005.

Ronell, Avital. *Dictations: On Haunted Writing*. 1986. Reprint, Urbana: University of Illinois Press, 2006.

Rostand, Jean. *La vie et ses problèmes*. Paris: Flammarion, 1939.

Rudy, Jason R. *Electric Meters: Victorian Physiological Poetics*. Athens: Ohio University Press, 2009.

Rzepka, Charles J. *Detective Fiction*. Cultural History of Literature. Cambridge: Polity, 2005.

Scheick, William J. "An Intrinsic Luminosity: Poe's Use of Platonic and Newtonian Optics." *Southern Literary Journal* 24, no. 2 (1992): 90–105.

Schelling, Friedrich Wilhelm Joseph von. *First Outline of a System of the Philosophy of Nature*. Translated by Keith R. Peterson. Albany: State University of New York Press, 2004.

Schleifer, Ronald. *Modernism and Time: The Logic of Abundance in Literature, Science, and Culture, 1880–1930*. Cambridge: Cambridge University Press, 2000.

Serres, Michel. *Feux et signaux de brume, Zola*. Paris: Grasset, 1975.

———. *Hermès IV: La distribution*. Paris: Éditions de Minuit, 1977.

Shelley, Mary Wollstonecraft. *Frankenstein*. Edited by J. Paul Hunter. Norton Critical Edition. New York: W. W. Norton, 1996.

Shelley, Percy Bysshe. *Essays, Letters from Abroad, Translations and Fragments*. Edited by Mary Wollstonecraft Shelley. Vol. 1. London: Edward Moxon, 1840.

Silk, Michael. "Metaphor and Metonymy: Aristotle, Jakobson, Ricoeur, and Others." In *Metaphor, Allegory, and the Classical Tradition: Ancient Thought and Modern Revisions*, edited by G. R. Boys-Stones, 115–47. Oxford: Oxford University Press, 2003.

Simon, Linda. *Dark Light: Electricity and Anxiety from the Telegraph to the X-Ray*. Orlando: Harcourt, 2004.

Simpson, Thomas K. *Figures of Thought: A Literary Appreciation of Maxwell's Treatise on Electricity and Magnetism*. Santa Fe: Green Lion Press, 2005.

Singer, George John. *Elements of Electricity and Electro-chemistry*. London: Longman, Hurst, Rees, Orme, Brown, and R. Triphook, 1814.

———. "Remarks on Some Electrical and Electrochemical Phenomena." *Journal of Natural Philosophy, Chemistry and the Arts* 31 (1812): 216–21.

Snyder, Laura J. *Reforming Philosophy: A Victorian Debate on Science and Society*. Chicago: University of Chicago Press, 2006.

Stamos, David N. *Edgar Allan Poe, Eureka, and Scientific Imagination*. Albany: State University of New York Press, 2017.

Steen, Gerard. "Metonymy Goes Cognitive-Linguistic." *Style* 39, no. 1 (2005): 1–11.

Steinle, Friedrich. "Experiments in History and Philosophy of Science." *Perspectives on Science* 10, no. 4 (2002): 408–32.

———. *Exploratory Experiments: Ampère, Faraday, and the Origins of Electrodynamics*. Translated by Alex Levine. Pittsburgh: University of Pittsburgh Press, 2016.

———. "Romantic Experiment? The Case of Electricity." Paper presented at Ciencia y Romanticismo, Maspalomas, Spain, 2002.

———. "Work, Finish, Publish? The Formation of the Second Series of Faraday's *Experimental Researches in Electricity*." *Physics* 33 (1996): 141–220.

Stern-Gillet, Suzanne. "On (Mis)interpreting Plato's *Ion*." *Phronesis* 49, no. 2 (2004): 169–201.

Sutton, Geoffrey. "Electric Medicine and Mesmerism." *Isis* 72, no. 3 (1981): 375–92.

Sweeney, Susan Elizabeth. "The Magnifying Glass: Spectacular Distance in Poe's 'Man of the Crowd' and Beyond." *Poe Studies* 36, no. 1 (2003): 3–17.

———. "Solving Mysteries in Poe, or Trying To." In *The Oxford Handbook of Edgar Allan Poe*, edited by J. Gerald Kennedy and Scott Peeples, 189–204. Oxford: Oxford University Press, 2019.

Swift, Jonathan. *Gulliver's Travels*. London: Penguin Classics, 2003.

Swinden, Jan Hendrik van. *Recueils de mémoires sur l'analogie de l'électricité et du magnétisme*. 3 vols. The Hague: Les libraires associés, 1784.

Thompson, Gary Richard. *Poe's Fiction: Romantic Irony in the Gothic Tales*. Madison: University of Wisconsin Press, 1973.

Thompson, Silvanus P. *Michael Faraday: His Life and Work*. London: Cassell, 1898.

Thouvenel, Pierre. *Mémoire physique et médicinal montrant des rapports évidens entre les phénomènes de la baguette divinatoire, du magnétisme et de l'électricité*. Paris: Didot le jeune, 1781.

Townshend, Chauncy Hare. *Facts in Mesmerism, with Reasons for a Dispassionate Inquiry into It*. London: Longman, Orme, Green, & Longman, 1840.

Trembley, Abraham. *Mémoires pour servir à l'histoire d'un genre de polypes d'eau douce, à bras en forme de cornes*. Leiden, Netherlands: J. & H. Verbeek, 1744.

Tresch, John. "Electromagnetic Alchemy in Balzac's *The Quest for the Absolute*." In *The Shape of Experiment*, edited by Henning Schmidgen and Julia Kursell, 57–77. Berlin: Max-Planck preprint, 2007.

———. "'Matter No More': Edgar Allan Poe and the Paradoxes of Materialism." *Critical Inquiry* 42, no. 4 (2016): 865–98.

———. *The Romantic Machine: Utopian Science and Technology After Napoleon*. Chicago: University of Chicago Press, 2012.

Valéry, Paul. "Le bilan de l'intelligence." In *Œuvres*. Bibliothèque de la pléiade, vol. 1, 1057–83. Paris: Gallimard, 1957.

Van Schlun, Betsy. *Science and the Imagination: Mesmerism, Media, and the Mind in Nineteenth-Century English and American Literature*. Berlin: Galda + Wilch Verlag, 2007.

Vermeir, Koen. "Athanasius Kircher's Magical Instruments: An Essay on 'Science,' 'Religion' and Applied Metaphysics." *Studies in History and Philosophy of Science* 38, no. 2 (2007): 363–400.

———. "Magnetic Theology as a Baroque Phenomenon." Paper presented at the Baroque Science Conference, Sydney, Australia, February 2008.

Villiers de L'Isle-Adam, Auguste. *Tomorrow's Eve*. Translated by Robert Martin Adams. Urbana: University of Illinois Press, 1982.

Voltaire. *Lettres philosophiques*. Paris: Flammarion, 2006.

Wallace, Richard. "'Amaze Your Friends!' Lucretius on Magnets." *Greece and Rome* 43, no. 2 (October 1996): 178–87.

Walls, Laura Dassow. "'Every Truth Tends to Become a Power': Emerson, Faraday, and the Minding of Matter." In *Emerson for the Twenty-First Century: Global Perspectives on an American Icon*, edited by Barry Tharaud, 301–17. Newark: University of Delaware Press, 2010.

Whewell, William. "Modern Science—Inductive Philosophy." Review of J. Hershel's *Preliminary Discourse on the Study of Natural Philosophy*. *Quarterly Review* 45 (1831): 374–407.

———. *Of Induction: With Especial Reference to Mr. J. Stuart Mill's System of Logic*. London: John W. Parker, 1849.

———. *The Philosophy of the Inductive Sciences: Founded Upon Their History*. Vol. 1. London: John W. Parker, 1840.

Willis, Martin. *Mesmerists, Monsters, and Machines: Science Fiction and the Cultures of Science in the Nineteenth Century*. Kent: Kent State University Press, 2006.

Wilson, Eric. *Emerson's Sublime Science*. New York: St. Martin's Press, 1999.

Winter, Alison. *Mesmerized: Powers of Mind in Victorian Britain*. Chicago: University of Chicago Press, 1998.

Wood, Gaby. *Edison's Eve: A Magical History of the Quest for Mechanical Life*. New York: Anchor Books, 2002.

Worth, Aaron. *Imperial Media: Colonial Networks and Information Technologies in the British Literary Imagination, 1857–1918*. Columbus: Ohio State University Press, 2014.

Zanetti, François. "Magnétisme animal et électricité médicale au dix-huitième siècle." In *Mesmer et mesmérismes: Le magnétisme animal en contexte*, edited by Bruno Belhoste and Nicole Edelman, 102–18. Paris: Omniscience, 2015.

Zwarg, Christina. "Vigorous Currents, Painful Archives: The Production of Affect and History in Poe's 'Tale of the Ragged Mountains.'" *Poe Studies* 43, no. 1 (2010): 7–33.

Index

Page numbers in *italics* refer to figures.

CPSIA information can be obtained
at www.ICGtesting.com
Printed in the USA
LVHW091952140420
653438LV00002B/2